STUDENT UNIT G

A2 Physics
UNIT 5

Edexcel

Module 5: Fields and Forces

Graham George

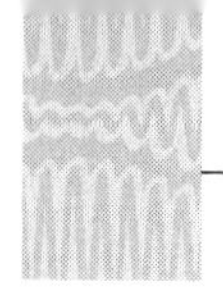

Philip Allan Updates
Market Place
Deddington
Oxfordshire
OX15 0SE

Orders
Bookpoint Ltd, 130 Milton Park, Abingdon, Oxfordshire, OX14 4SB
tel: 01235 827720
fax: 01235 400454
e-mail: uk.orders@bookpoint.co.uk
Lines are open 9.00 a.m.–5.00 p.m., Monday to Saturday, with a 24-hour message answering service. You can also order through the Philip Allan Updates website: www.philipallan.co.uk

ISBN-13: 978-1-84489-571-7
ISBN-10: 1-84489-571-8

This guide has been written specifically to support students preparing for the Edexcel A2 Physics Unit 5 examination. The content has been neither approved nor endorsed by Edexcel and remains the sole responsibility of the author. Exam questions are reproduced by permission of Edexcel Ltd. Edexcel accepts no responsibility whatsoever for the accuracy or method of working in the answers given.

Printed by MPG Books, Bodmin

P00764

Contents

Introduction

About this guide

This guide is one of a series covering the Edexcel specification for AS and A2 physics. It offers advice for the effective study of **Unit 5: Fields and Forces**. Unit 5 also includes the Practical Test, but this guide is concerned only with the 'Fields and Forces' part of the unit. The Practical Test is covered in the guide for Unit 6.

The aim of this guide is to help you to *understand* the physics — it is not intended as a shopping list, enabling you to cram for an examination. The guide has three sections:

- **Introduction** — this gives brief guidance on approaches and techniques to ensure you answer the examination questions in the best way that you can.
- **Content Guidance** — this section is not intended to be a detailed textbook. It offers guidance on the main areas of the content of Unit 5, with an emphasis on worked examples. These examples illustrate the type of questions that you are likely to come across in the examination.
- **Questions and Answers** — this comprises two unit tests, compiled from recent past papers, to give the widest possible coverage of the unit content. Answers are provided; in some cases, distinction is made between responses that might have been given by an A-grade candidate and those typical of a C-grade candidate. Common errors made by candidates are also highlighted so that you, hopefully, do not make the same mistakes.

The effective understanding of physics requires time and effort. No one suggests it is an easy subject, but even those who find it difficult can overcome their problems by the proper investment of time.

The development of an understanding of physics can only evolve with experience, which means time spent thinking about physics, working with it and solving problems. This book provides you with the platform to do this. If you try all the worked examples and the unit tests *before* looking at the answers (no cheating!), you will begin to think for yourself as well as develop the necessary techniques for answering examination questions. In addition, you will need to *learn* the basic formulae, definitions and experiments. Thus prepared, you will be able to approach the examination with confidence.

The specification

The specification states the physics that will be examined in the unit tests and describes the format of those tests. This is not necessarily the same as what teachers might choose to teach or what you might choose to learn.

The purpose of this book is to help you with Unit Test 5, but don't forget that what you are doing is learning *physics*. The specification can be obtained from Edexcel, either as a printed document or from the web at **www.edexcel.org.uk**

The unit test

The examination

The assessment of Unit 5 is made up of two components, with an equal weighting given to each. The Practical Test is taken at a separate time from the Unit Test and is discussed fully in the guide for Unit 6. Unit Test 5, which lasts just 1 hour, usually consists of five compulsory structured questions, each allocated from 4 to 12 marks. The total number of marks for the paper is 40, which counts for 15% of the A2 or 7.5% of the total A-level mark. Questions will assume that Units 1, 2 and 4 have been studied, but will not examine the content again in detail. For example, a knowledge of forces and Newton's laws (Unit 1), together with circular motion (Unit 4), will be expected in the study of satellites, while the basics of circuit electricity (Unit 2) will be required when discussing electric and magnetic fields.

The test will examine assessment objectives AO1 (knowledge with understanding) and AO2 (application of knowledge and understanding, synthesis and evaluation), giving an equal weighting to each objective. This is a slight change of emphasis towards AO2 compared with AS papers. The questions are structured, but less so than at AS. In particular, quantities may not be given in base units so, for example, you may have to convert measurements in μA and kΩ into 10^{-6} A and 10^{3} Ω respectively before substituting into the appropriate equation.

Command terms

Examiners use certain words that require you to respond in a particular way. You must distinguish between these terms and understand exactly what each requires you to do. Some frequently used commands are shown below.

- **State** — a brief sentence giving the essential facts; no explanation is required (nor should you give one).
- **Define** — you can use a *word equation*, but if you use *symbols*, you must state what each symbol represents.
- **List** — simply a series of words or terms; there is no need to write sentences.
- **Outline** — a logical series of bullet points or phrases will suffice.
- **Describe** — a diagram is essential for an experiment, then give the main points concisely (again, bullet points can be used).
- **Draw** — diagrams should be drawn in section, neatly and *fully labelled* with all measurements clearly shown, but don't waste time — remember it is not an art exam.
- **Sketch** — usually a graph, but graph paper is not necessary (although a grid is sometimes provided); however, the axes must be labelled, including a scale if

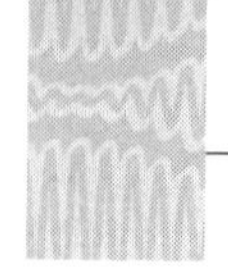

numerical data are given, the origin should be shown if appropriate and the general shape of the expected line should be drawn.

- **Explain** — use correct physics terminology and principles; the depth of your answer should reflect the number of marks available.
- **Show that** — usually a value is given so that you can proceed with the next part; you should show all your working and give your answer to more significant figures than the value given (to prove you have actually done the calculation).
- **Calculate** — show all your working and give *units* at every stage; the number of significant figures in your answer should reflect the given data, but you should keep each stage in your calculator to prevent excessive rounding.
- **Determine** — means you will probably have to extract some data, often from a graph, in order to perform a calculation.
- **Estimate** — a calculation in which you have to make a sensible assumption, possibly about the value of one of the quantities — think, does this give a reasonable answer?
- **Suggest** — there is often no single correct answer; credit is given for sensible reasoning based on correct physics.
- **Discuss** — you need to sustain an argument, giving evidence before and against, based on your knowledge of physics, possibly using appropriate data to justify your answer.

You should pay particular attention to diagrams, sketching graphs and calculations. Candidates often lose marks by failing to label diagrams properly, by not giving essential numerical data on sketch graphs and by not showing all their working, or by omitting units, in calculations.

As in the other unit guides, quantity algebra is used in the solutions to worked examples and in the answers to the Unit Tests. This involves the use of the appropriate unit with *every* physical quantity in an equation (not just in the answer). Although this is not specifically required in the examination, it is good practice to use quantity algebra as it makes you think carefully about the use of units and will help you eliminate silly mistakes.

Revision

The purpose of this introduction is not to provide you with a detailed guide to revision techniques — there are many books written on study skills if you feel you need more help in preparing for examinations. There are, however, some points worth mentioning that will help you when revising for your physics A-level:

- Be familiar with what you need to know — ask your teacher and look through the specification.
- Make sure you have a good set of notes — you cannot revise properly from a textbook.
- Learn *all* the equations indicated in the specification and be familiar with the equations that are provided for you in the examination (at the end of each question paper), so that you can find them quickly and use them correctly.

- Make sure you learn definitions thoroughly and in detail, e.g. in the definition of magnetic field strength using the formula $F = BIl$, the conductor must be *perpendicular to the field*.
- Be able to describe (with a diagram) the basic experiments referred to in the specification.
- Make revision active by writing out equations and definitions, drawing diagrams, describing experiments and by performing lots of calculations.

Content Guidance

This section is a guide to the content of **Unit 5: Fields and Forces**. It does not constitute a textbook for Unit 5 material.

The main areas of this unit are:

- Gravitational fields — gravitational field strength; Newton's law of gravitation for the force between point masses; radial and uniform fields; equipotentials; application to satellites
- Electric fields — electrostatic phenomena, electric charge and its measurement, conductors and insulators, the discrete nature of the electronic charge; Coulomb's law for the force between point charges; electric field strength in radial and uniform fields; equipotentials and electric potential difference; electron beams
- Magnetic fields — permanent magnets, field lines, concept of a neutral point; magnetic flux density and its measurement for the field of a long straight wire and for that of a solenoid; force on a current-carrying conductor in a magnetic field
- Electromagnetic induction — magnetic flux linkage; Faraday's and Lenz's laws; the transformer

There are many similarities between the three types of field, but also some differences. In particular, in a gravitational field, masses *always attract*, whereas in electric and magnetic fields, we can have both *attraction* and *repulsion*. You need to develop an understanding of these similarities and differences as you consider each type of field.

The study of gravitational fields involves forces and circular motion. You should therefore familiarise yourself with these topics from Units 1 and 4. In the same way, the development of electric and magnetic fields depends on the knowledge and understanding of the basic electricity covered in Unit 2. You may, therefore, find it helpful to look back over this unit as well, to remind yourself of its contents.

You should also consult a standard textbook for more information. Physics is a logical discipline and you need to have a feeling for where the information you are dealing with fits into the broader pattern of the subject. Remember that the specification tells you merely what can be examined in the unit test. Only by wider reading will you be able to understand how the different aspects of physics all come together to help us understand the wonders of the universe in which we live.

Gravitational fields

In physics, a **field** is defined as a **region in which a body experiences a force**. This unit looks at three types of field: gravitational, electric and magnetic.

A **gravitational field** is a region in which a mass experiences a force due to another mass. The field with which you are familiar is, of course, the one that we all live in — the Earth's gravitational field. We have a mass, which experiences a force due to the Earth's gravitational field — we call this force our **weight**. The Earth also has a magnetic field and an electric field, which will be discussed later.

Gravitational field strength

The symbol for gravitational field strength is g, which is also used for the acceleration due to gravity. Gravitational field strength is defined as:

$$g = \frac{F}{m}$$

where F is the force acting on a mass, m, placed in the field. Strictly speaking, m should be a *point mass* so that it does not affect the field. From the equation we can see that the units of gravitational field strength are **N kg^{-1}**. The gravitational field strength at the Earth's surface has a value with which you are already familiar, namely 9.81 N kg^{-1}.

Worked example

Show that the units for the acceleration due to gravity, m s^{-2}, are the same as those for the gravitational field strength, N kg^{-1}.

Answer

From $F = ma$, N ≡ kg m s^{-2}

As $g = \frac{F}{m}$, the units of $g \equiv \frac{\text{kg m s}^{-2}}{\text{kg}} \equiv \text{m s}^{-2}$ ≡ units of acceleration

Force between point masses

From his work on gravitation, **Newton** showed that all bodies attract each other with a gravitational force that is proportional to each of their masses and inversely proportional to the square of their distance apart. This is written as:

$$F = G\frac{m_1 m_2}{r^2}$$

where m_1 and m_2 are *point* masses a distance r apart, and G is the universal gravitational constant. G is 'universal' as it applies to all masses throughout the universe.

Gravitational forces are very weak, unless very large masses are involved, and so G has a very small value:

$$G = 6.67 \times 10^{-11} \text{ N m}^2 \text{ kg}^{-2}$$

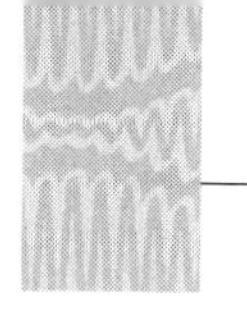

Although the equation is specifically for point masses, it can also be applied to two uniform spheres, in which case r is the distance apart of the *centres* of the spheres. **Newton's law of gravitation**, as it is called, can therefore be applied to a good approximation to the Sun and its planets, including the Earth and the Moon.

Worked example

Estimate the gravitational force of attraction between:

(a) two students, each of mass 75 kg, standing next to each other

(b) two oil tankers passing each other a distance of 50 m apart (the oil tankers each have a mass of 500 000 tonnes and are 60 m wide)

State any assumptions that you make in arriving at your answers.

Answer

(a) Assuming the students can be considered to be point masses a distance of, say, 0.6 m apart, then:

$$F = G\frac{m_1 m_2}{r^2}$$

$$= 6.67 \times 10^{-11}\ \text{N m}^2\ \text{kg}^{-2} \times \frac{75\ \text{kg} \times 75\ \text{kg}}{0.60^2\ \text{m}^2} \approx 1 \times 10^{-6}\ \text{N}$$

(b) Assuming that all the mass of each tanker acts at its centre, the distance between the centres when they pass each other is 50 m + (2 × 30) m = 110 m. You will also need to know that 1 tonne = 1000 kg. Then:

$$F = 6.67 \times 10^{-11}\ \text{N m}^2\ \text{kg}^{-2} \times \frac{5 \times 10^8\ \text{kg} \times 5 \times 10^8\ \text{kg}}{110^2\ \text{m}^2} \approx 1\ \text{kN}$$

The Earth's gravitational field

A gravitational field is represented by **lines of force**. These lines of force indicate the direction in which a mass would move in the field. As gravitational field strength has a *direction* associated with it, it is a **vector** quantity. A strong field is represented by drawing the lines of force close together and a uniform field by drawing the lines parallel and equally spaced.

A spherically symmetrical mass, m, has a **radial** gravitational field. The gravitational field strength, g, due to this mass is inversely proportional to the *square* of the distance, r, from the *centre* of the mass and is given by:

$$g = G\frac{m}{r^2}$$

As the Earth is approximately spherical, its gravitational field is radial (diagram (a)). However, within a small region close to the Earth's surface, the scale is such that we can consider the field to be **uniform** (diagram (b)).

(a)

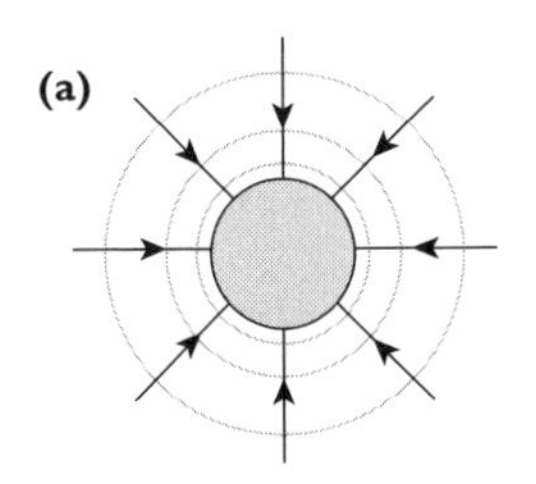

(b)

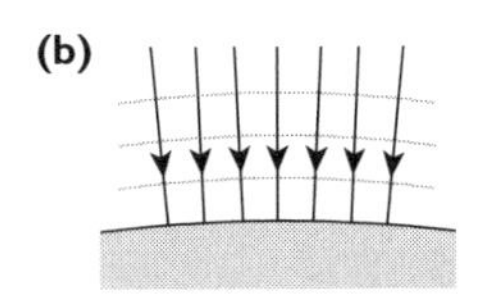

Worked example

(a) Given that the Earth has a radius of 6400 km and the gravitational field strength at its surface is 9.8 N kg^{-1}, show that the mass of the Earth is about 6×10^{24} kg.

(b) Estimate a value for the Earth's gravitational field strength at the edge of the Earth's atmosphere, which you may take to be 600 km above the Earth's surface.

Answer

(a) From $g = G\dfrac{m}{r^2}$ we get

$$m = \frac{gr^2}{G} = \frac{9.8\ \text{N kg}^{-1} \times (6400 \times 10^3)^2\ \text{m}^2}{6.67 \times 10^{-11}\ \text{N m}^2\ \text{kg}^{-2}} = 6.0 \times 10^{24}\ \text{kg}$$

(b) If the edge of the atmosphere is 600 km above the Earth's *surface*, then it is a distance of 600 + 6400 = 7000 km from the Earth's *centre*, giving:

$$g = G\frac{m}{r^2} = 6.67 \times 10^{-11}\ \text{N m}^2\ \text{kg}^{-2} \times \frac{6.0 \times 10^{24}\ \text{kg}}{(7000 \times 10^3)^2\ \text{m}^2} = 8.2\ \text{N kg}^{-1}$$

A mass in a gravitational field will have a force acting on it and so it has the possibility of doing work if it is allowed to move in the direction of the field. It is said to have **gravitational potential**. An **equipotential surface** joins up points in a gravitational field that have the same potential. Some equipotentials, in two dimensions, are shown by the lighter lines in diagrams (a) and (b). Note that they are equally spaced in the uniform field of diagram (b), but get further apart in the radial field of diagram (a). They are like *contour lines* on a map, which join together points of equal height. Equipotentials are always *at right angles* to the lines of force.

In a uniform field, where g (and hence the gravitational force) remains constant, the work done by a mass, m, falling through a height, Δh, is given by:

$$\Delta W = F \times \text{distance moved in direction of force}$$
$$= mg\Delta h$$

Remember, g is only constant *close* to the Earth's surface (see the worked example above, and the graph on p. 14) and so this relationship holds only near the Earth.

As g is inversely proportional to r^2, the variation of g with distance from the Earth can be expressed graphically as shown in the following graph.

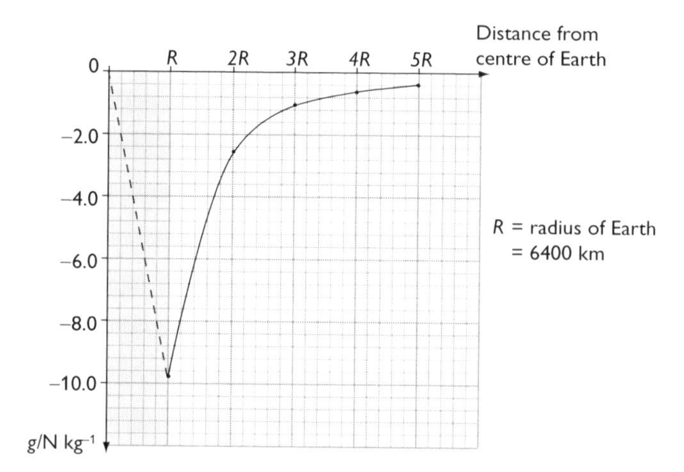

Note that g is a *vector* quantity. As it acts in a direction *towards* the Earth it is shown as being *negative* in the above graph, in which the distance away from the Earth's centre is taken as positive.

Satellites

For a satellite to remain in orbit round the Earth, it needs a *resultant* force to act on it, which is directed *towards the centre* of the orbit. You should be familiar with this concept from studying circular motion in Unit 4. In the case of a satellite, this force is the gravitational attraction of the Earth. As this force is towards the centre of the Earth, the centre of the orbit is the centre of the Earth.

In the examination, you are likely to meet problems about three types of satellite:

- those having orbits very close to the Earth (used for research into the Earth's atmosphere)
- 'geostationary' communications satellites, which orbit the Earth with the same *angular velocity* as the rotation of the Earth, thus remaining above the same point of the Earth
- the Earth's own satellite, the Moon

In all cases, the assumption is made that the orbits are circular, which may not actually be true in reality.

Worked example

A satellite orbits the Earth, radius R_E and mass M_E, in a circular path of radius r with angular speed ω, as shown in the diagram.

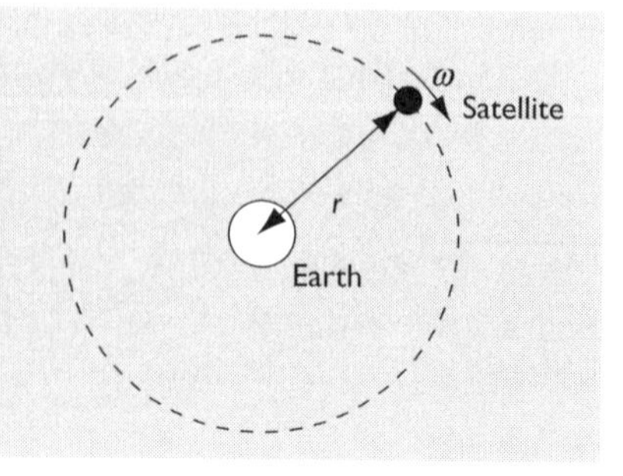

(a) By considering the force acting on a mass m on the Earth's surface, show that:

$$gR_E^2 = GM_E$$

(b) Hence show for the satellite that $\omega^2 = \frac{gR_E^2}{r^3}$

(c) A geostationary satellite has a period $T = 24$ hours. Determine its height above the Earth, assuming that $R_E = 6400$ km.

Answer

(a) At the Earth's surface $F = mg = \frac{GM_E m}{R_E^2}$, giving $gR_E^2 = GM_E$

(b) Resultant force on satellite $= m\omega^2 r = \frac{GM_E m}{r^2}$ which gives $\omega^2 = \frac{GM_E}{r^3} = \frac{gR_E^2}{r^3}$

(c) As $\omega = \frac{2\pi}{T}$ we get $\omega^2 = \frac{4\pi^2}{T^2} = \frac{gR_E^2}{r^3}$ so $r^3 = \frac{gR_E^2 T^2}{4\pi^2}$

$$r^3 = \frac{9.81\ \text{N kg}^{-1} \times (6400 \times 10^3)^2\ \text{m}^2 \times (24 \times 60 \times 60)^2\ \text{s}^2}{4\pi^2}$$

$r = 4.24 \times 10^7$ m

But this is the radius of orbit from the Earth's centre, so we must subtract the Earth's radius to find the height above the Earth:

height $= 4.24 \times 10^7$ m $- 6400 \times 10^3$ m $= 3.6 \times 10^7$ m (= 36 000 km)

You must remember that the Moon has its own gravitational field (of strength about one-sixth of that of the Earth at its surface), so there is a point somewhere between the Earth and the Moon where their fields are equal and opposite, giving a resultant field of zero.

Worked example

The diagram shows a body of mass m situated at a point which is a distance R from the centre of the Earth and r from the centre of the moon.

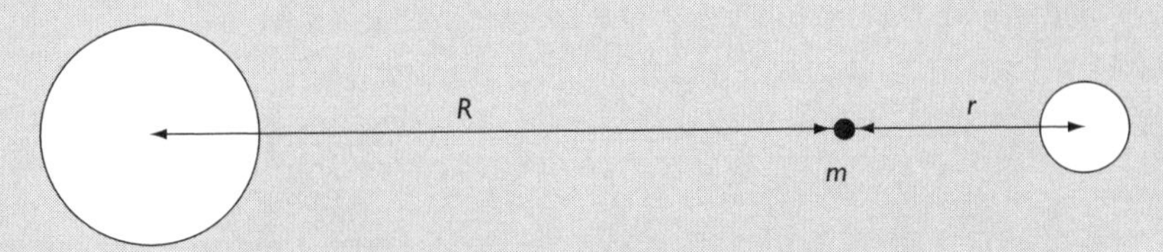

(a) The masses of the Earth and Moon are M_E and M_M respectively. The universal gravitational constant is G. Using the symbols given, write down an expression for:

(i) the gravitational force of attraction between the body and the Earth

(ii) the gravitational force of attraction between the body and the Moon

(b) Determine the value of R for which the resultant gravitational force on this body is zero if $M_E = 81M_M$ and the mean distance from the Earth to the Moon is 380 000 km.

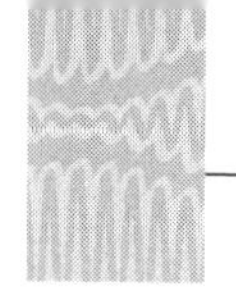

Answer

(a) (i) $F_E = \dfrac{GM_E m}{R^2}$

(ii) $F_M = \dfrac{GM_M m}{r^2}$

(b) If resultant force is zero, $\dfrac{GM_E m}{R^2} = \dfrac{GM_M m}{r^2}$

giving $\dfrac{M_E}{M_M} = \dfrac{R^2}{r^2} = 81$, so $R = 9r$

$R + r = 380\,000$ km. This means that $R = 0.9 \times 380\,000$ km $= 340\,000$ km.

Electric fields

Electrostatic phenomena

Can you explain why, when a balloon is rubbed on your hair, it can then mysteriously be made to stick to a wall?

When the balloon is rubbed, the work you do against friction transfers some of the electrons from your hair to the balloon, which therefore becomes negatively charged, leaving your hair positively charged. You will notice that your hair stands up. This is because each hair is positively charged and is repelled by the hair next to it.

When you put the balloon on the wall, electrons at the surface of the wall are repelled to the other side of their atoms by the negative charge on the balloon. The positive surface of the wall will then attract the negative charges on the balloon, which remains stuck to the wall.

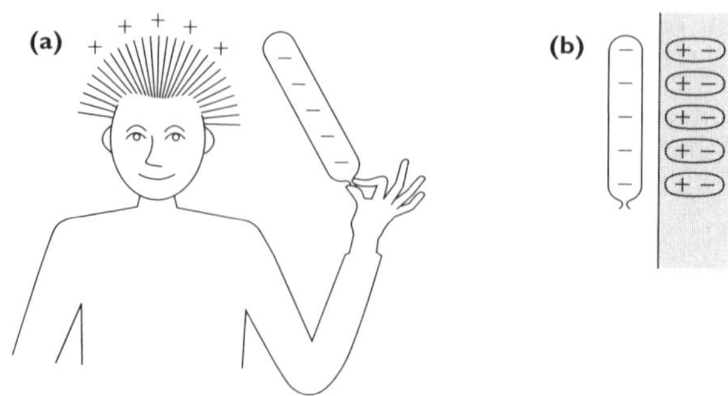

This simple experiment shows that there are *two types of charge*, which we call positive and negative, and that *like charges attract* whereas *unlike charges repel*. Charging the balloon is called *charging by friction* and the process by which the balloon sticks to the wall is called *charging by induction*. You should remember that in all cases it is the *electrons* that move, leaving behind positively charged atoms in the region they have moved from.

When the balloon is charged, it is negative with respect to the Earth, but the charge does not flow to Earth through your body even though your body is a *conductor*. This is because the rubber in the balloon is an *insulator*. A conductor allows charge to flow in it when there is a potential difference whereas an insulator will not.

You need to understand all the above terms that are in italics. You should also understand the *discrete*, or *quantum*, nature of the electronic charge. The smallest charge that can exist as a free entity is that of a single electron, which is 1.6×10^{-19} C. All other charges are whole number multiples of this charge. Charge can be measured by an electronic device called a *coulombmeter*, just like an ammeter measures current. You may already have seen a coulombmeter when investigating the photoelectric effect in Unit 4.

Worked example

In an investigation of the photoelectric effect, a negatively charged zinc plate is connected to a coulombmeter, which registers 8 nC. A student accidentally touches the zinc plate, which discharges to Earth in a time of 4 ms.

(a) How many electrons are there in a charge of 8 nC?

(b) What is the average current in the student when the plate discharges?

Answer

(a) Number of electrons $= \dfrac{8 \times 10^{-9}\ \text{C}}{1.6 \times 10^{-19}\ \text{C}} = 5 \times 10^{10}$ electrons

(b) $I = \dfrac{\Delta Q}{\Delta t}$, so $I = \dfrac{8 \times 10^{-9}\ \text{C}}{4 \times 10^{-3}\ \text{s}} = 2\ \mu\text{A}$

Electric field strength

An electric field is a region in which an electric charge experiences a force due to another charge. **Electric field strength**, symbol E, is defined by the equation:

$$E = \frac{F}{Q}$$

where F is the force experienced by a charge Q placed in the field. The charge should theoretically be a small, point charge so that it does not affect the field.

Like gravitational field strength, electric field strength is a *vector*. Its direction is defined as the direction in which a *positive charge* would move if placed in the field. From the above equation, we can see that the units of electric field strength are N C^{-1}, although we will see later that we can also use V m^{-1}.

Worked example

The Earth has an electric field strength of about 120 N C^{-1} at its surface, in a direction *towards* the Earth. A speck of dust of mass 1.0×10^{-18} g carries a negative charge equal to that of one electron. Ignoring any upthrust on the speck of dust due to the air:

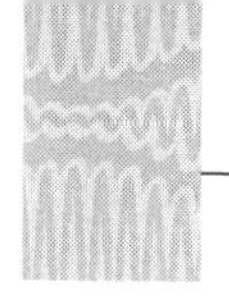

(a) Draw a free-body force diagram for the speck of dust.

(b) Calculate:

(i) the weight of the speck of dust

(ii) the electrostatic force on the speck of dust

(iii) the resultant force on the speck of dust

Answer

(a)

Electrostatic force (EQ)

Speck of dust

Weight (mg)

(b) (i) Weight $W = mg = 1.0 \times 10^{-18}$ kg $\times$ 9.8 N kg^{-1} = 9.8×10^{-18} N

(ii) From $E = \dfrac{F}{Q}$ we get $F = EQ = 120$ N C^{-1} $\times 1.6 \times 10^{-19}$ C
$= 1.92 \times 10^{-17}$ N

(iii) Resultant force = 1.92×10^{-17} N upwards − 9.8×10^{-18} N downwards
= 9.4×10^{-18} N *upwards*

Force between point charges

The electric force between charges is just like the gravitational force between masses. Coulomb first showed that the force between two charges was proportional to each of the charges and inversely proportional to the square of their distance apart. For two *point* charges Q_1 and Q_2 a distance r apart *in free space*, the force F between them is given by:

$$F = k\frac{Q_1Q_2}{r^2}$$

where k is the Coulomb law constant, having the value $k = \dfrac{1}{4\pi\varepsilon_0} = 9.0 \times 10^9$ N m^2 C^{-2}.

The constant ε is a property of the material in which the field is situated, called its **electrical permittivity**. For free space it is assigned the symbol ε_0 and has the value 8.85×10^{-12} F m^{-1}. To a very good approximation ε for air has the same value. Examination questions will usually involve the use of the constant k.

The force between charges differs from that between masses as it can be either attractive (the force between unlike charges) or repulsive (the force between like charges), whereas the force between masses is *always* attractive. As the constant k is very large (~10^{10}) compared with G, which is extremely small (~10^{-10}), electric forces tend to be much larger than gravitational forces, even when the charges are small (~μC).

Worked example

Two identical table tennis balls, A and B, each of mass 1.5 g are attached to insulating threads. Both the balls are charged to the same positive value. When the threads are fastened to a point P the balls hang as shown in the diagram.

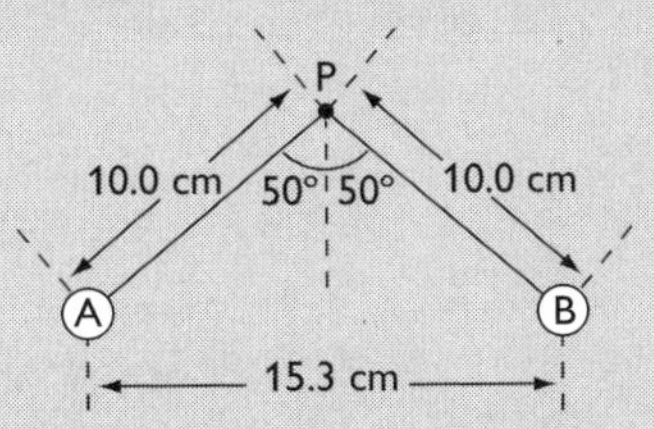

(a) Draw a labelled free-body force diagram for ball A.

(b) Calculate the tension in each of the threads.

(c) Show that the electrostatic force between the two balls is about 2×10^{-2} N.

(d) Calculate the charge on each ball.

(e) How does the gravitational force between the two balls compare with the electrostatic force in part (c)?

Answer

(a)

Pull of string (tension)

Electrical force

A

Pull of earth (weight)

(b) $T \cos 50° = mg$, so $T = \dfrac{mg}{\cos 50°} = \dfrac{1.5 \times 10^{-3}\text{ kg} \times 9.81\text{ N kg}^{-1}}{\cos 50°} = 2.3 \times 10^{-2}\text{ N}$

(c) $F_E = T \sin 50° = 2.3 \times 10^{-2}\text{ N} \times \sin 50° = 1.75 \times 10^{-2}\text{ N} \approx 2 \times 10^{-2}\text{ N}$

(d) $F_E = \dfrac{kQ_1Q_2}{r^2} = \dfrac{kQ^2}{r^2}$ where Q is the charge on each ball

$$Q^2 = \frac{F_E r^2}{k} = \frac{1.75 \times 10^{-2}\text{ N} \times (15.3 \times 10^{-2})^2\text{ m}^2}{9.0 \times 10^{9}\text{ N m}^2\text{ C}^{-2}} = 4.68 \times 10^{-14}\text{ C}^2$$

$Q = 2.1 \times 10^{-7}$ C

(e) The gravitational force is *very much smaller* than the electrostatic force and it is *attractive*.

Radial and uniform electric fields

A point charge (or a charged spherical conductor) produces a **radial** field which, like the gravitational field of a point mass, obeys an inverse square law:

$$E = \frac{kQ}{r^2}$$

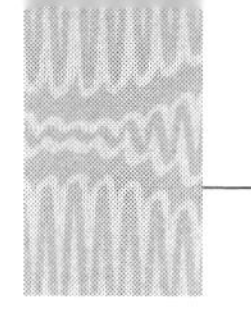

As before, the shape of the field is represented by **lines of force** which indicate the direction in which a *positive charge* would move if placed in the field. Similarly, **equipotential surfaces** join points in the field that are at the *same electric potential*. A radial field is shown in diagram (a), with (two-dimensional) equipotentials shown as lighter lines. Note that, as before, the equipotentials are *at right angles* to the lines of force.

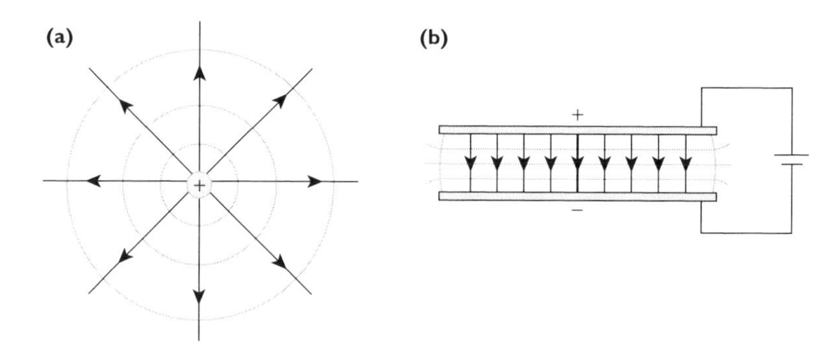

A *uniform* electric field can be created by applying a potential difference between two parallel conducting plates. The field loses its uniformity near either end of the plates. This is shown in diagram (b). Note that the equipotentials are equally spaced in a uniform field. It can be shown that for a uniform field:

$$E = \frac{V}{d}$$

where V is the potential difference between two equipotentials that are a distance d apart. For the parallel plate arrangement, V is the potential difference between the plates, and d is their distance apart.

From $E = V/d$ we have an alternative unit for electric field strength, namely V m^{-1}.

Worked example 1

Show that the two units used for electric field strength, N C^{-1} and V m^{-1}, are equivalent.

Answer

From $V = \frac{W}{Q}$, the units for V = J C^{-1}

and we also know that from work done J = N m

This gives us V m^{-1} = J C^{-1} m^{-1} = N m C^{-1} m^{-1} = N C^{-1}

(The equation $V = W/Q$ comes from Unit 2, and is also discussed below.)

Worked example 2

An isolated charged metallic sphere is brought close to an identical earthed sphere that initially was uncharged. Part of the resultant electric field and some of its equipotential lines are shown in the diagram, which is drawn to scale.

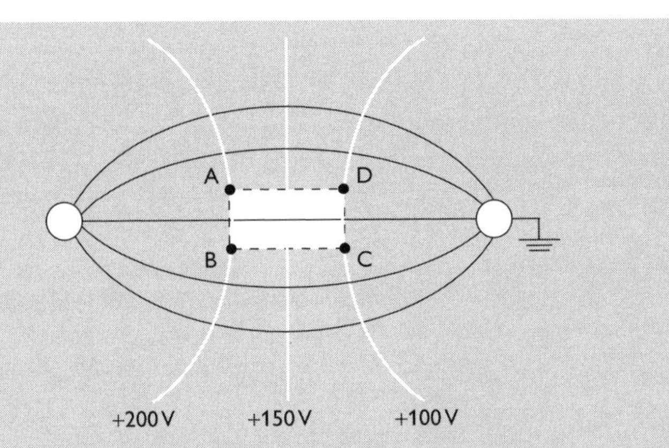

(a) On the diagram mark the sign of the charge on each sphere.

(b) Draw arrows on the field lines to indicate the direction of the electric field.

(c) Estimate the electric field strength in the region ABCD. What assumption do you have to make?

Answer

(a) The left-hand sphere is *positive* and the right-hand sphere is *negative*.

(b) The arrows on each of the five field lines are from *left* to *right* (i.e. + to –).

(c) Assuming that the field is *approximately uniform* in the region ABCD

$$E = \frac{V}{d} \approx \frac{(200 - 100)\text{ V}}{14 \times 10^{-3}\text{ m}} \approx 7000 \text{ V m}^{-1}$$

Work done in an electric field

In Unit 2 electric potential difference was defined as $V = W/Q$, so the work done moving a charge Q through a potential difference V is $W = QV$. If V is in volts and Q is in coulombs, then W will be in joules.

When an electron (charge 1.6×10^{-19} C) is moved through a potential difference of 1 V, the work done on it is $W = QV = 1.6 \times 10^{-19}\text{ C} \times 1\text{ V} = 1.6 \times 10^{-19}$ J. This is a convenient quantity of energy when dealing with electrons and is called an **electronvolt** (eV):

$$1 \text{ eV} = 1.6 \times 10^{-19} \text{ J}$$

Worked example 1

With reference to the diagram above, calculate how much work is done against the field in taking an electron from:

(a) A to B

(b) A to C

Give your answers in both eV and J.

Answer

(a) As A and B are at the *same potential*, ***no*** work is done against the field in taking an electron from A to B.

(b) $V_{AC} = (200 - 100)\ \text{V} = 100\ \text{V}$
$W = QV = 100\ \text{eV} = 100\ \text{V} \times 1.6 \times 10^{-19}\ \text{C} = 1.6 \times 10^{-17}\ \text{J}$

Worked example 2

The diagram shows two parallel metal plates with a potential difference of 3000 V applied across them. The plates are in a vacuum.

(a) Copy the diagram and sketch the electric field pattern in the region between the plates.

(b) On the same diagram sketch and label two equipotential lines.

(c) The plates are 25 mm apart. Show that the force experienced by an electron just above the bottom plate is about 2×10^{-14} N.

(d) Sketch a graph to show how the force on the electron varies with the distance of the electron from the bottom plate.

(e) This force causes the electron to accelerate. The electron is initially at rest in contact with the bottom plate when the potential difference is applied. Calculate its speed as it reaches the top plate.

Answer

(a) and **(b)**

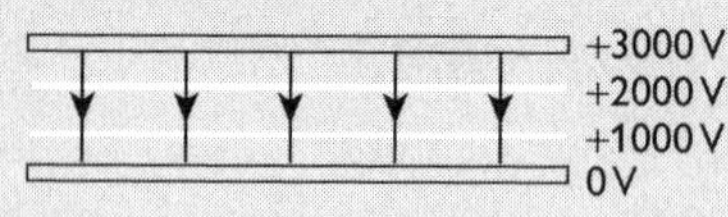

(c) $E = \dfrac{V}{d} = \dfrac{3000\ \text{V}}{25 \times 10^{-3}\ \text{m}} = 1.2 \times 10^{5}\ \text{V m}^{-1}$

$E = \dfrac{F}{Q}$ so $F = EQ = 1.2 \times 10^{5}\ \text{V m}^{-1} \times 1.6 \times 10^{-19}\ \text{C} = 1.9(2) \times 10^{-14}\ \text{N}$

(d)

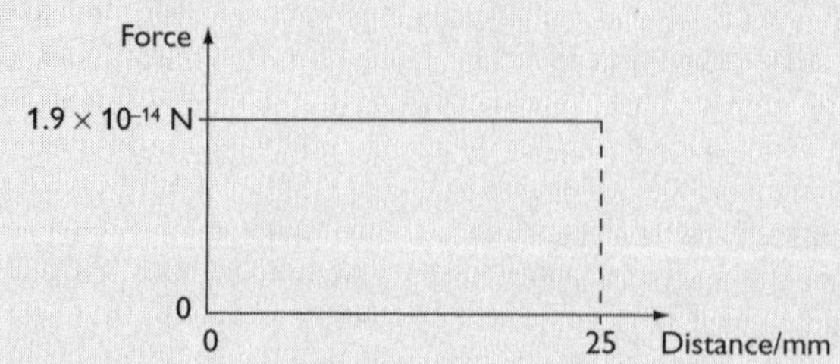

(e) $F = ma$ so $a = \dfrac{F}{m} = \dfrac{1.92 \times 10^{-14}\ \text{N}}{9.11 \times 10^{-31}\ \text{kg}}$

$a = 2.11 \times 10^{16}$ m s^{-2}
$v^2 - u^2 = 2as$ (where $u = 0$ as electron starts from rest)
$v^2 - 0^2 = 2 \times 2.11 \times 10^{16}$ m s$^{-2} \times 25 \times 10^{-3}$ m
$v = 3.2 \times 10^7$ m s^{-1}

Electron beams

Another, and quicker, way of approaching part (e) above is by equating the work done on the electron by the electric field to the kinetic energy gained by the electron. If we let the charge on the electron be e, we get:

$$\Delta\left(\frac{1}{2}m_e v^2\right) = e\Delta V$$

Worked example
What is the speed of an electron after it has been accelerated through 3000 V (as in the previous example)?

Answer

$$\Delta\left(\frac{1}{2}m_e v^2\right) = e\Delta V \text{ so } v^2 \frac{2e\Delta V}{m_e}$$

$$= \frac{2 \times 1.6 \times 10^{-19}\text{ C} \times 3000\text{ V}}{9.11 \times 10^{-31}\text{ kg}}$$

$v = 3.2 \times 10^7$ m s^{-1} (as before)

Because electrons (and other charged particles, such as alpha particles) have such a small mass, they undergo extremely large accelerations in even relatively small electric fields. They can therefore acquire very high speeds, approaching the speed of light; in which case the simple equation above is no longer applicable and we have to use relativistic mechanics — beyond the scope of A-level. In air, ionisation would take place, slowing down the particles, so examples assume that the particles are in a vacuum.

Worked example
The diagram shows a high-speed alpha particle entering the space between two charged plates in a vacuum.

(a) Copy the diagram and draw the subsequent path of the alpha particle as it passes between the plates and well beyond them.

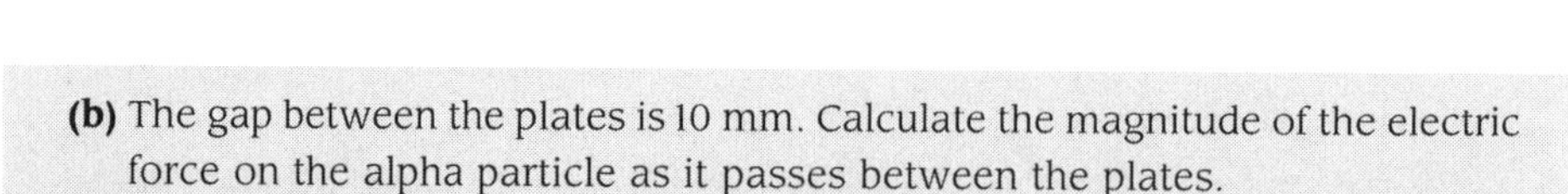

(b) The gap between the plates is 10 mm. Calculate the magnitude of the electric force on the alpha particle as it passes between the plates.

Answer

(a)

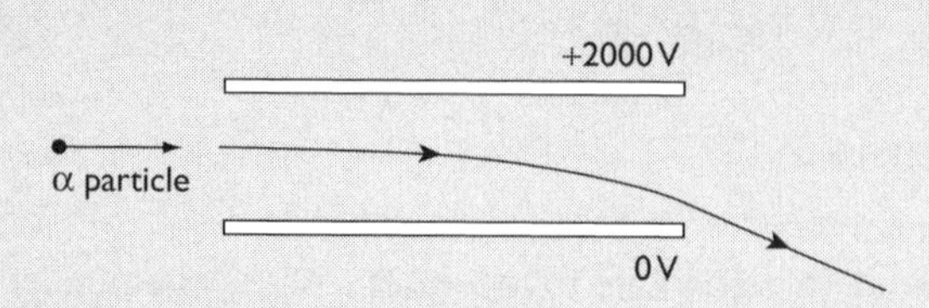

Note that the path is parabolic when the alpha particle is between the plates. As soon as it leaves the plates, there is no longer any force acting on it and so, by Newton's first law, it continues in a *straight line*.

(b) $E = \frac{V}{d} = \frac{2000\text{ V}}{10 \times 10^{-3}\text{ m}} = 2.0 \times 10^5\text{ V m}^{-1}$

$F = EQ = 2.0 \times 10^5\text{ V m}^{-1} \times 2 \times 1.6 \times 10^{-19}\text{ C}$
$= 6.4 \times 10^{-14}\text{ N}$

Note that we need two electronic charges as this is an alpha particle.

Capacitance

Capacitors

A capacitor is an electrical device that can *store charge*. Although, like resistors, capacitors are manufactured, any electrical conductor, from a tin can to the Earth, can store charge and so can be considered a capacitor.

The ability of a capacitor to store charge is called its **capacitance** (C) and is defined by:

$$C = \frac{Q}{V}$$

where Q is the charge stored when the capacitor is charged to a potential V.

The units of capacitance are therefore C V^{-1}. (Be careful not to be confused by *italic C* for capacitance and upright C for coulombs.) As capacitance is a common quantity, it is given its own unit, the farad (F). A farad is an extremely large value, so we usually use the microfarad ($\mu\text{F} = 10^{-6}$ F) or even the picofarad ($\text{pF} = 10^{-12}$ F) in practical situations. You should also be aware of the fact that capacitors usually have a large manufacturing **tolerance**, typically as much as 20%. This means that the manufacturer only guarantees the value of the capacitor to within 20% of its **nominal** (stated) value. This is important to remember in practical work.

Worked example

The following circuit is set up to investigate the charge stored on a capacitor.

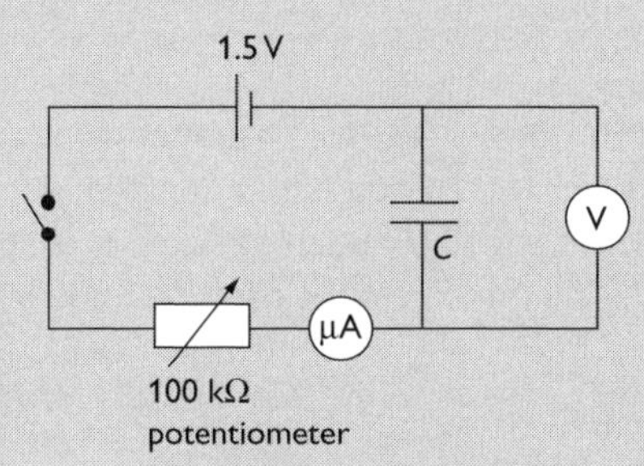

The switch is closed and the stopwatch is started. The current is maintained at 15 μA by continually adjusting the potentiometer. The potential difference V across the capacitor is recorded every 30 s for 3 minutes, as shown in the table below.

t/s	V/V	Q/
30	0.21	
60	0.41	
90	0.62	
120	0.83	
150	1.04	
180	1.25	

(a) Complete the table by adding values of Q, with an appropriate unit.

(b) Plot a graph of Q against V.

(c) Use your graph to determine a value for the capacitance of the capacitor.

Answer

(a) From Unit 2, $Q = It = 15\ \mu A \times 30\ s = 450\ \mu C$ (900 μC, 1350 μC,...,2700 μC)

(b)

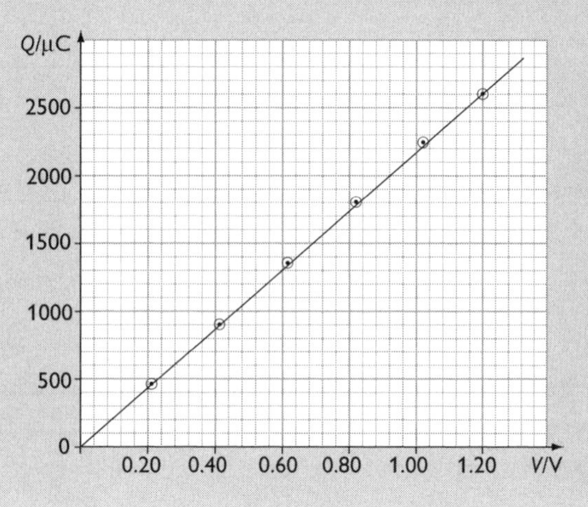

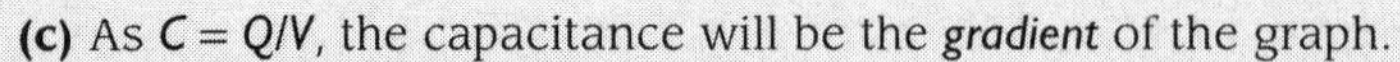

(c) As $C = Q/V$, the capacitance will be the ***gradient*** of the graph.

$$C = \frac{2600\ \mu C}{1.20\ V} = 2200\ \mu F$$

If a capacitor is charged and then discharged through a resistor, a graph of the form shown below is obtained. This is in fact an *exponential* curve (like radioactive decay), but you do not need to know this until Unit 6. What you *do* need to know is that the *area* under the curve is equal to the *charge* lost by the capacitor (from $Q = It$).

Worked example

The circuit below is used to investigate the discharge of a capacitor.

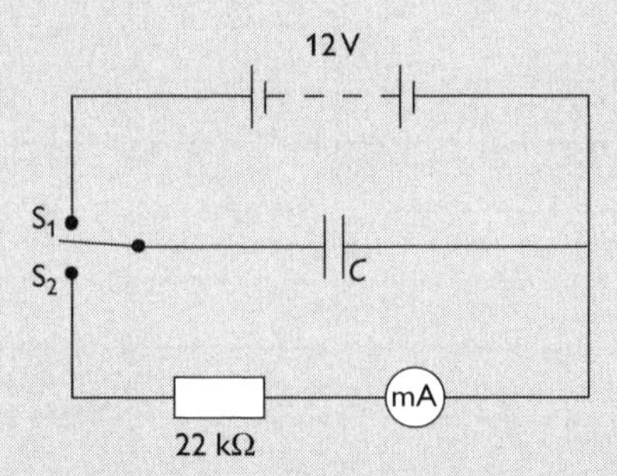

With the switch in position S_1 the capacitor is charged. The switch is then moved to S_2 and readings of current and time are taken as the capacitor discharges through the resistor. The results are plotted on the graph below.

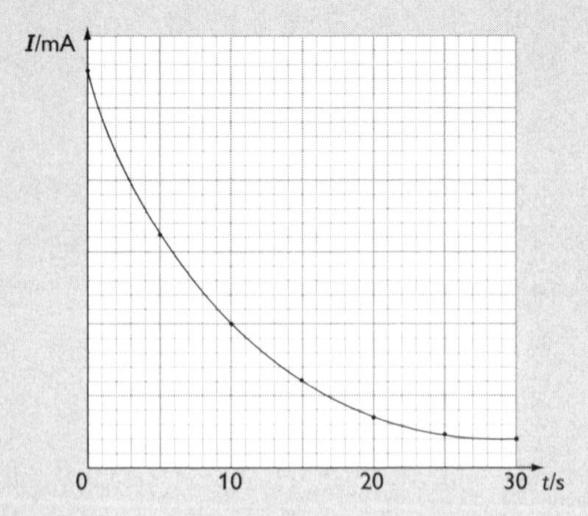

(a) Calculate the current at the instant switch S_2 is closed and hence add a scale to the y-axis.

(b) Estimate the charge stored on the capacitor before it is discharged. Explain any assumptions that you made in arriving at your answer.

(c) Hence determine an approximate value for the capacitance of the capacitor.

Answer

(a) At $t = 0$, $I = \frac{V}{R} = \frac{12\text{ V}}{22 \times 10^3\ \Omega} = 0.55$ mA

A scale of 0.1 mA per large division can now be added to the y-axis.

(b) The amount of charge discharged after 30 s will be the area under the curve up to this point. Although the capacitor is not quite fully discharged after 30 s, this amount of charge will be almost as much as the charge that is initially stored in the capacitor.

There are approximately 11 large squares. Each of these is equivalent to a charge of 0.1 mA × 5 s = 0.5 mC, giving the charge stored $Q \approx 11 \times 0.5$ mC ≈ 5.5 mC.

(c) $C = \frac{Q}{V} \approx \frac{5.5 \times 10^{-3}\text{ C}}{12\text{ V}} \approx 4.6 \times 10^{-4}\text{ F} \approx 460\ \mu\text{F}$

Capacitors in series and parallel

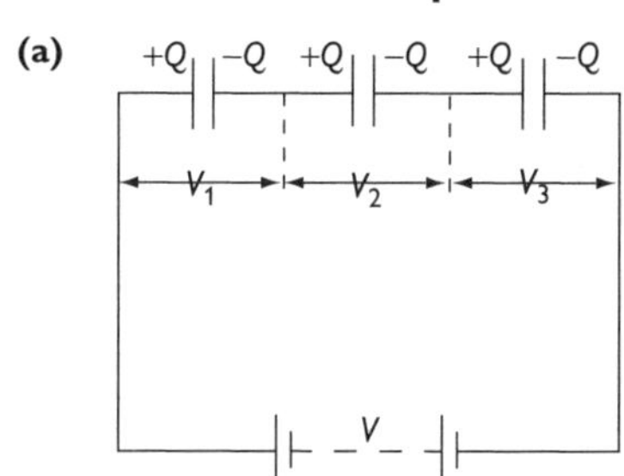

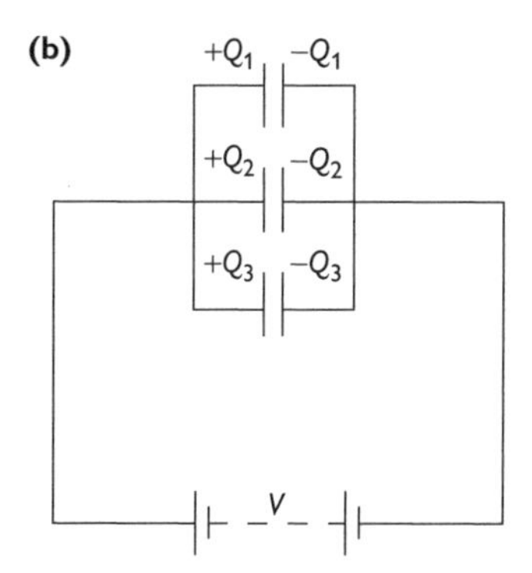

In *series* (diagram (a)), each capacitor has the *same charge* (as they charge by induction) and also $V = V_1 + V_2 + V_3$. From $C = Q/V$ we get $V = Q/C$, so:

$\frac{Q}{C} = \frac{Q}{C_1} + \frac{Q}{C_2} + \frac{Q}{C_3}$, which gives $\frac{1}{C} = \frac{1}{C_1} + \frac{1}{C_2} + \frac{1}{C_3}$

In *parallel* (diagram (b)), each capacitor has the *same potential* across it and the total charge is $Q = Q_1 + Q_2 + Q_3$. From $C = Q/V$ we get $Q = CV$, so:

$CV = C_1V + C_2V + C_3V$ so $C = C_1 + C_2 + C_3$

These formulae must not be confused with the similar looking formulae for *resistance*. Remember, resistors in series simply add up, whereas in parallel they add up reciprocally. For capacitors it is the *other way round*.

Worked example

(a) A 100 μF capacitor is connected to a 12 V supply.

(i) Calculate the charge stored.

(ii) Show on the diagram the arrangement and magnitude of charge on the capacitor.

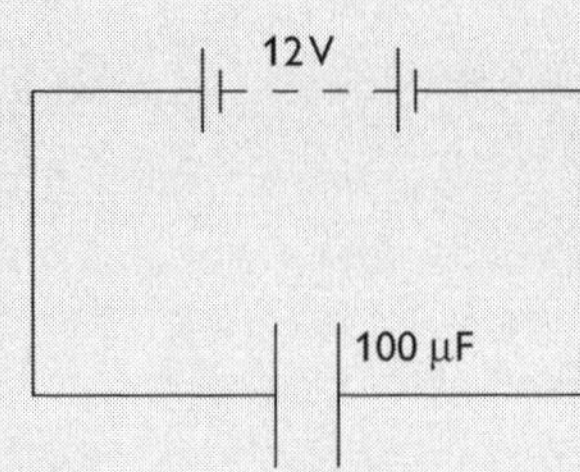

(b) This 100 μF charged capacitor is disconnected from the battery and is then connected across a 400 μF uncharged capacitor.

(i) What happens to the charge initially stored on the 100 μF capacitor?

(ii) Calculate the new voltage across the capacitors.

(iii) How much charge is stored on each capacitor?

(c) The capacitors are now disconnected and discharged. They are then reconnected in series with the 12 V supply. Calculate:

(i) the charge on each capacitor

(ii) the potential across each capacitor

Answer

(a) (i) The charge stored is $Q = CV = 100 \times 10^{-6}\ \text{F} \times 12\ \text{V} = 1200\ \mu\text{C}$

(ii)

12 V

100 μF

+1200 μC   −1200 μC

(b) (i) Some of the charge on the 100 μF capacitor will transfer to the 400 μF capacitor.

(ii) The equivalent capacitance $C = 100\ \mu\text{F} + 400\ \mu\text{F} = 500\ \mu\text{F}$ (as they are in *parallel*).

New voltage $V = \dfrac{Q}{C} = \dfrac{1200 \times 10^{-6}\ \text{C}}{500 \times 10^{-6}\ \text{F}} = 2.4\ \text{V}$

(iii) From $Q = CV$

$Q_{100} = 100 \times 10^{-6}\ \text{F} \times 2.4\ \text{V} = 240\ \mu\text{C}$

$Q_{400} = 400 \times 10^{-6}\ \text{F} \times 2.4\ \text{V} = 960\ \mu\text{C}$

(Note that the total charge adds up to the original 1200 μC.)

(c) (i) For capacitors in *series* $\frac{1}{C} = \frac{1}{C_1} + \frac{1}{C_2} + \frac{1}{C_3}$, giving

$$\frac{1}{C} = \frac{1}{100\ \mu\text{F}} + \frac{1}{400\ \mu\text{F}} = \frac{4+1}{400\ \mu\text{F}} = \frac{5}{400\ \mu\text{F}} \text{ so } C = \frac{400\ \mu\text{F}}{5} = 80\ \mu\text{F}$$

Charge on each capacitor $Q = CV = 80 \times 10^{-6}\ \text{F} \times 12\ \text{V} = 960\ \mu\text{C}$

(ii) $V_{100} = \frac{Q}{C} = \frac{960 \times 10^{-6}\ \text{C}}{100 \times 10^{-6}\ \text{F}} = 9.6\ \text{V}$

$V_{400} = \frac{960 \times 10^{-6}\ \text{C}}{400 \times 10^{-6}\ \text{F}} = 2.4\ \text{V}$

(Note that the voltages add up to the 12 V of the supply.)

Energy stored in a charged capacitor

From our definition that $V = W/Q$ we have $W = VQ$. If, therefore, we plot a V–Q graph for charging a capacitor, the area under the graph represents the work done charging the capacitor. This will then be the energy stored in the capacitor.

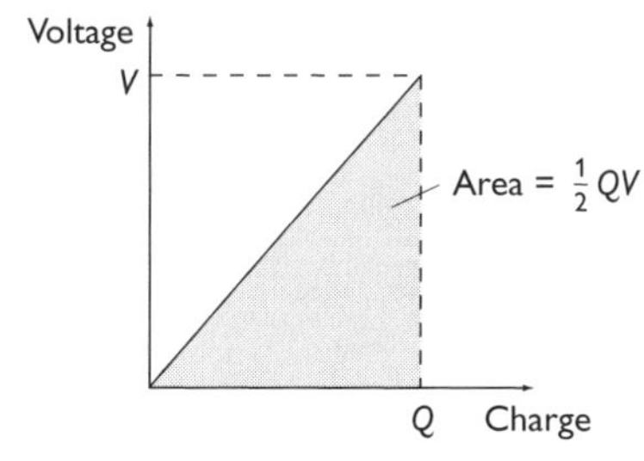

From the diagram we can see that the energy given by the shaded area is $\frac{1}{2}QV$. We usually want to know how much energy is stored when we charge a capacitor of known capacitance C to a potential V. If we substitute $Q = CV$ into the above equation we get:

$$W = \frac{1}{2}CV^2$$

Worked example

A defibrillator is a machine that is used to correct irregular heartbeats by passing a large current through the heart for a short time. The machine uses a 6000 V supply to charge a capacitor of capacitance 20 μF. The capacitor is then discharged through the metal electrodes (defibrillator paddles) which have been placed on the chest of the patient.

(a) (i) Calculate the charge on the capacitor plates when charged to 6000 V.

(ii) Calculate the energy stored in the capacitor.

When the capacitor is discharged, there is an initial current of 40 A through the patient.

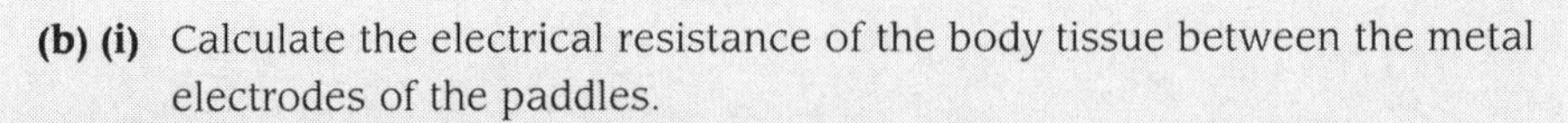

(b) (i) Calculate the electrical resistance of the body tissue between the metal electrodes of the paddles.

(ii) Assuming a constant discharge rate of 40 A, calculate how long it would take to discharge the capacitor.

(iii) In practice, the time for discharge is longer than this calculated time. Suggest a reason for this.

Answer

(a) (i) Charge $Q = CV = 20 \times 10^{-6}\text{ F} \times 6000\text{ V} = 0.12\text{ C}$

(ii) Energy stored $W = \frac{1}{2}CV^2 = \frac{1}{2} \times 20 \times 10^{-6}\text{ F} \times 6000^2\text{ V}^2 = 360\text{ J}$

(b) (i) Resistance $R = \frac{V}{I} = \frac{6000\text{ V}}{40\text{ A}} = 150\ \Omega$

(ii) From $Q = It$, we get $t = \frac{Q}{I} = \frac{0.12\text{ C}}{40\text{ A}} = 3.0 \times 10^{-3}\text{ s} = 3.0\text{ ms}$

(iii) In reality the time will be longer than this because as the capacitor discharges, the potential gets less and so the current will also get less, causing a decrease in the rate of discharge.

Magnetic fields

Permanent magnets

You should already be familiar with the magnetic field of a bar magnet. Magnetic fields are described by field lines, just like gravitational and electric fields. In the case of a magnetic field, a field line indicates the direction in which a *north pole* would move. As *like poles repel* and *unlike poles attract*, a magnetic field line will run from the north pole of a magnet to the south pole. Some magnetic fields are shown below.

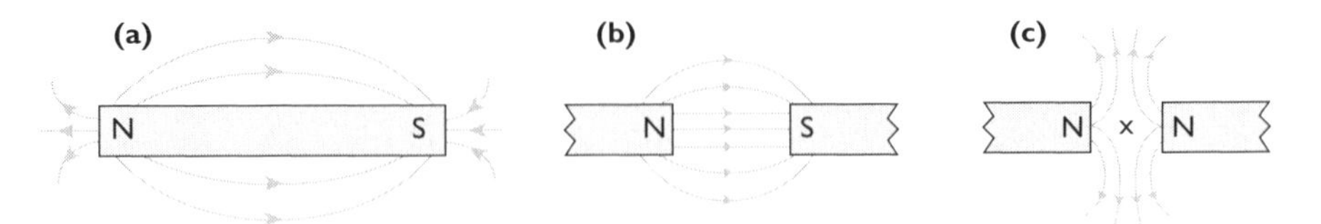

A region in a magnetic field where two fields are equal and opposite, and therefore give a resultant field of zero, is called a **neutral point**. This is indicated by an × in diagram (c).

A magnetic field is caused by a *moving electric charge*, usually a current in a conductor. In the case of permanent magnets, their fields are caused by *electrons spinning* on their axes in the atoms of the magnetic material.

Magnetic flux density

The strength of a magnetic field is called its **magnetic flux density** and is given the symbol B. It is determined by the *force* the field exerts on an electric current and is therefore a **vector** quantity. Its *magnitude* is defined by:

$$F = BIl$$

where F is the force exerted by the field on a length l of conductor, *perpendicular to the field*, in which the current is I.

Rearranging, we get $B = \frac{F}{Il}$, giving the unit of B as N A^{-1} m^{-1}, called a **tesla** (T).

The *direction* of a magnetic field is given by Fleming's left-hand rule — if your **f**irst finger points in the direction of the **f**ield and your se**c**ond finger in the direction of the **c**urrent, then the force will cause the conductor to **m**ove in the direction in which your thu**m**b is pointing. This is shown in the diagram. When the switch is closed, the wire jumps up out of the magnet.

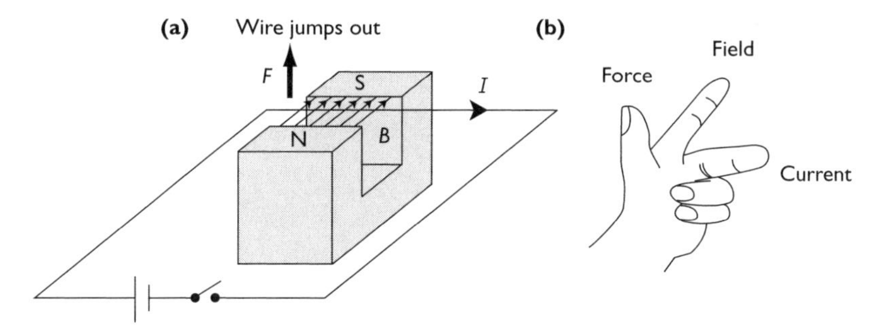

Just as it has a gravitational field and an electric field, the Earth also has a *magnetic field*. However, unlike the other fields, the Earth's magnetic field varies considerably from place to place, both in magnitude and in direction. It is also continuously changing — indeed, it has even completely reversed many times in the billions of years of the Earth's existence.

In London, in 2006, the Earth's magnetic flux density was about 48 μT and acted *downwards* at an angle of about 67° (diagram (a)).

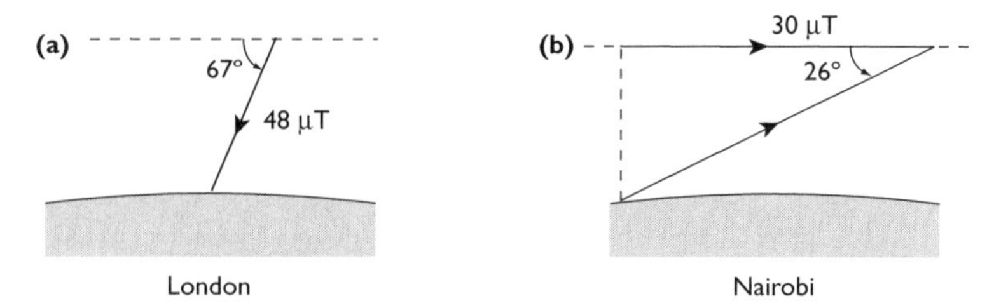

As we conduct most experiments in a horizontal plane, we are more interested in the *horizontal component* of the Earth's magnetic field. From diagram (a) above we can see this is given by:

$$B_H = 48\ \mu T \times \cos 67° = 19\ \mu T$$

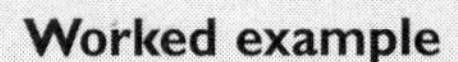

Worked example

In 2006, the horizontal component of the Earth's magnetic field in Nairobi was 30 μT, with the total flux density acting *upwards* at an angle of 26° to the horizontal (diagram (b) on p. 31). Calculate:

(a) the total flux density of the Earth's field

(b) the vertical component of the Earth's field

Answer

(a) $B_H = B \cos 26° \text{ so } B = \frac{B_H}{\cos 26°} = \frac{30\ \mu\text{T}}{\cos 26°} = 33\ \mu\text{T}$

(b) $\tan 26° = \frac{B_V}{B_H} \text{ so } B_V = B_H \tan 26° = 30\ \mu\text{T} \times \tan 26° = 15\ \mu\text{T}$ *upwards*

To demonstrate that the force on a conductor in a magnetic field is proportional to the current in the conductor we can use the arrangement shown in the diagram below.

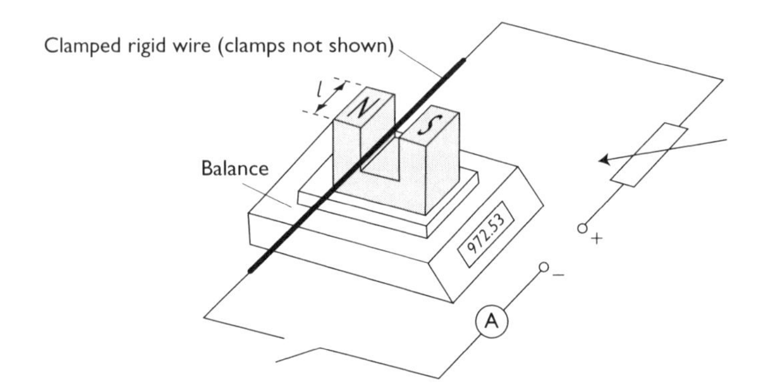

The reading on the balance is recorded. The switch is then closed so that there is a current I in the wire. The magnetic field exerts a force F on the wire, and by Newton's third law, the wire exerts an *equal and opposite* force F on the field. This causes the reading of the balance to increase. If this increase is Δm, then $F = \Delta m \times g$. The experiment is repeated with further values of current, and a graph of F against I is plotted. A straight line through the origin shows that the force on the wire is proportional to the current. To a good approximation, the field does not extend much beyond the length l of the pole pieces, which means an estimate for the field strength B can be found from $F = BIl$.

Worked example

An aluminium rod of mass 50 g is placed across two parallel horizontal copper tubes which are connected to a low-voltage supply. The aluminium rod lies across the centre of and perpendicular to the uniform magnetic field of a permanent magnet as shown in the diagram. The magnetic field acts over a region measuring 6.0 cm × 5.0 cm.

The magnetic flux density of the field between the poles is 0.20 T *into* the paper and the current in the rod is 4.5 A. Calculate:

(a) the magnitude and direction of the force acting on the aluminium rod

(b) the initial acceleration of the rod, assuming that it slides without rolling

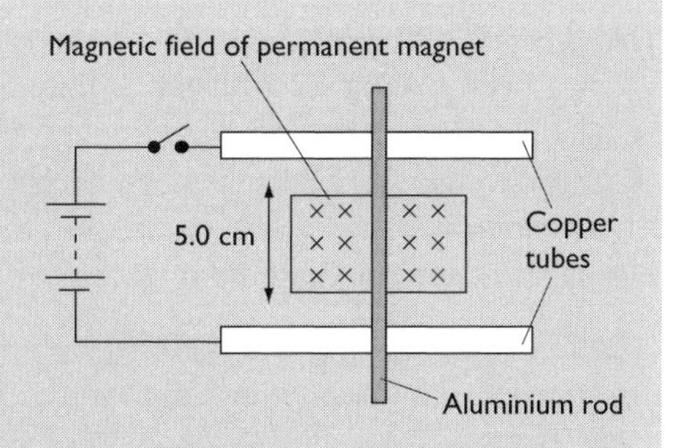

Answer

(a) The force is given by $F = BIl = 0.20\text{ T} \times 4.5\text{ A} \times 5.0 \times 10^{-2}\text{ m} = 0.045\text{ N}$

From the left-hand rule the direction of the force is to the *right*. (Point your first finger (field) *into* the paper, twist your hand clockwise so that your second finger (current) is pointing *down* the rod (towards you) and your thumb (force) should be pointing to the *right*.)

(b) From $F = ma$ we get $a = \dfrac{F}{m} = \dfrac{0.045\text{ N}}{50 \times 10^{-3}\text{ kg}} = 0.90\text{ m s}^{-2}$

Magnetic effect of a steady current

We saw earlier that a magnetic field is produced whenever we have a moving charge. A particular form of charge movement is the current in a wire. The magnetic field round a long **straight wire** in which there is a steady current is shown on the right.

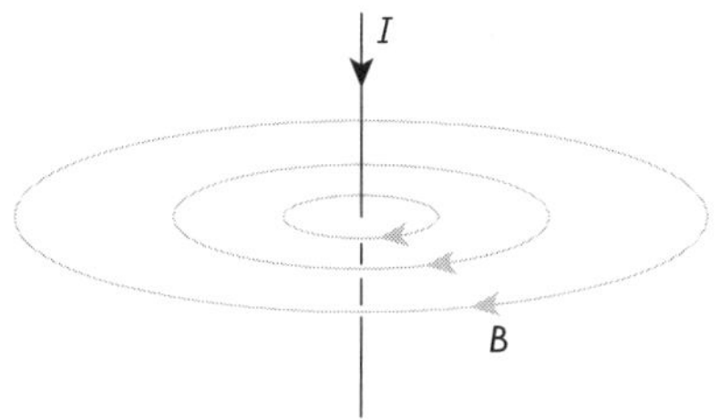

The *magnitude* of the field B a distance r from a wire carrying a current I is given by:

$$B = \frac{\mu_0 I}{2\pi r}$$

where μ_0 is a constant, called the **magnetic permeability of free space**. This is equivalent to G for a gravitational field and ε_0 for an electric field. From the definition of the ampere, which is one of the base SI units, μ_0 has a defined value of exactly $4\pi \times 10^{-7}\text{ N A}^{-2}$.

Worked example

Show that the unit of the permeability of free space, μ_0, is N A^{-2}.

Answer

From $F = BIl$ we get $B = \dfrac{F}{Il} = \dfrac{\text{N}}{\text{A m}} = \text{N A}^{-1}\text{ m}^{-1}$

From $B = \dfrac{\mu_0 I}{2\pi r}$ we get $\mu_0 = \dfrac{2\pi r B}{I} = \dfrac{\text{m N A}^{-1}\text{ m}^{-1}}{\text{A}} = \text{N A}^{-2}$

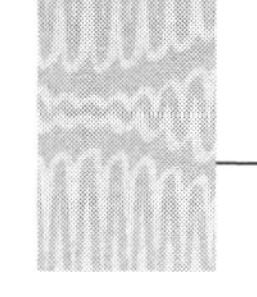

The *direction* of the magnetic field near a straight wire is given by the *screwdriver rule* — imagine screwing a screw *into* a piece of wood:

- The direction of the screw going *in* is the direction of the *current*.
- The direction in which your hand rotates, i.e. *clockwise*, is the direction of the *magnetic field*.

This is shown in the diagram on p. 33.

If you are unscrewing the screw (i.e. the current is coming out towards you), the direction of rotation of your hand (and the magnetic field) will be anticlockwise.

Worked example

Diagram (a) shows a straight wire Y carrying a current I.

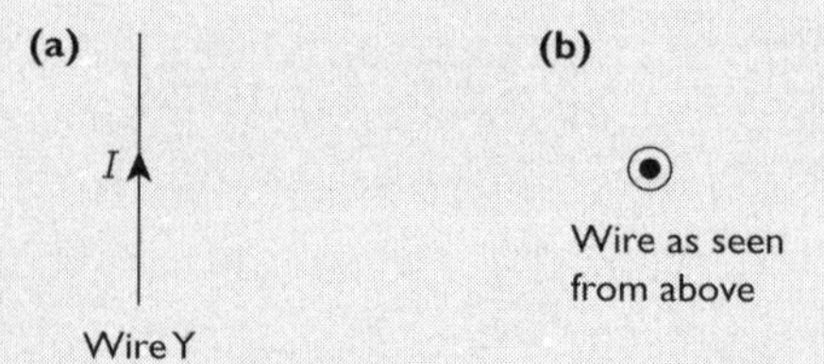

(a) (i) Copy diagram (b) and draw the magnetic field pattern close to the wire as seen looking from above.

(ii) Calculate the current in this wire when the field strength due to the wire alone at a point 12 cm from the centre of the wire is 1.4×10^{-5} T.

A second wire Z, carrying a current of $2I$, is placed 12 cm from wire Y.

The next diagram shows the direction of the force F_Z exerted on the second wire Z.

(b) Add an arrow to the diagram showing the direction of the current $2I$ in the second wire Z.

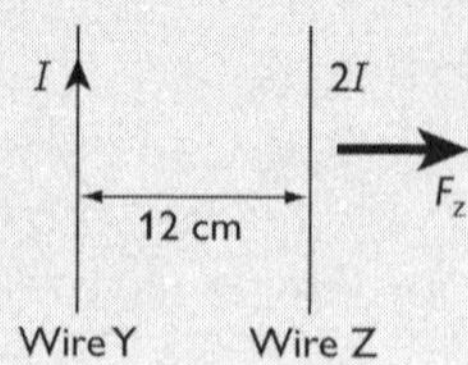

A force F_Y is also exerted on the first wire Y.

(c) (i) Add another arrow to the same diagram, showing the direction of this force. Label it F_Y.

(ii) What is the ratio F_Y:F_Z of the two forces?

(iii) State two methods by which the magnitude of the force F_Y could be reduced.

Answer

(a) (i)

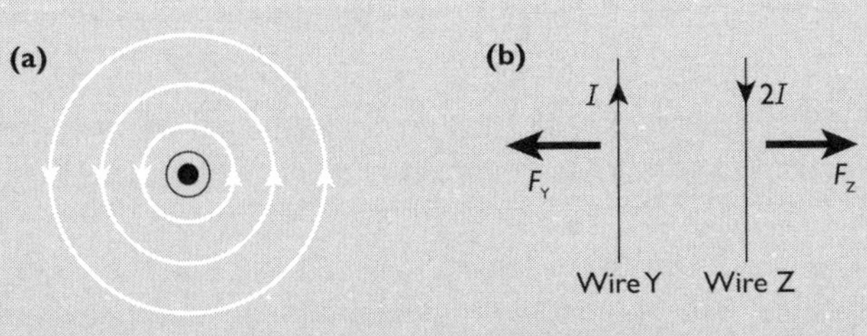

Note that the field lines should *not* be equally spaced, but should get *further apart*. The *direction* of the field should also be shown.

(ii) From $B = \dfrac{\mu_0 I}{2\pi r}$ we get $I = \dfrac{2\pi r B}{\mu_0} = \dfrac{2\pi \times 12 \times 10^{-2}\ \text{m} \times 1.4 \times 10^{-5}\ \text{T}}{4\pi \times 10^{-7}\ \text{N A}^{-2}} = 8.4\ \text{A}$

(b) In the region between the two wires, the field due to wire Y (anticlockwise looking from the top) is going *into the paper*. For the wires to *repel*, the field due to wire Z must be in the same direction, also *into* the paper. Looking from the top, this will be *clockwise* for wire Z, so the current in wire Z must be *downwards*.

(c) (i) By Newton's third law, F_Y is *equal* and *opposite* to F_Z and so F_Y will act on wire Y in the opposite direction to F_Z acting on wire Z, i.e. towards the left (diagram (b)).

(ii) From the above, the ratio F_Y:F_Z is 1:1.

(iii) The force F_Y could be reduced by

- reducing the current in Y or Z (or both)
- increasing the separation r
- reducing the length of wire Y

You also need to be familiar with the magnetic field of a **solenoid**, which is effectively a length of wire wound onto a tube. The field pattern is shown below in diagram (a). Outside the solenoid, the field pattern looks similar to that of a bar magnet (diagram (b)).

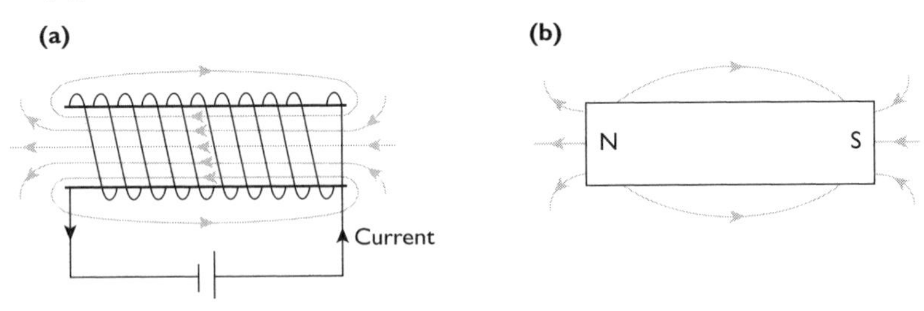

By comparing the field of the solenoid with that of the bar magnet, we can see that the end of the solenoid in which the current is anticlockwise can be considered to be a north pole and the end in which the current is clockwise as a south pole. One way of remembering this is as shown in the diagram on p. 36.

If the solenoid is *long* (typically 10 times longer than its diameter), the field in the central region of the solenoid is *uniform* and is of field strength:

$$B = \mu_0 nI$$

where I is the current in the solenoid and n is the number of turns *per unit length*. The field drops off near the ends of the solenoid, becoming exactly half this value at each end.

Worked example

A steel slinky spring is stretched so that its coils are 10 mm apart. The slinky is arranged horizontally on the bench so that its axis is in the east–west direction and a small plotting compass is placed at its centre. The slinky is then connected in series with a DC supply and an ammeter. The current in the slinky is 0.24 A.

(a) Show that the magnetic field at the centre of the slinky created by this current is about 30 μT.

(b) Through what angle will the compass deflect when the current is switched on if the horizontal component of the Earth's magnetic field is 19 μT?

Answer

(a) If the coils are 10 mm apart, there will be 100 coils in a 1 metre length, i.e. $n = 100\ \text{m}^{-1}$.
From $B = \mu_0 nI$, $B = 4\pi \times 10^{-7}\ \text{N A}^{-2} \times 100\ \text{m}^{-1} \times 0.24\ \text{A} = 30.2\ \mu\text{T} \approx 30\ \mu\text{T}$

(b) This field is along the axis of the slinky, i.e. east–west. There will also be the 19 μT horizontal component of the Earth's field acting north–south. By *vector addition* the resultant field will be as shown below:

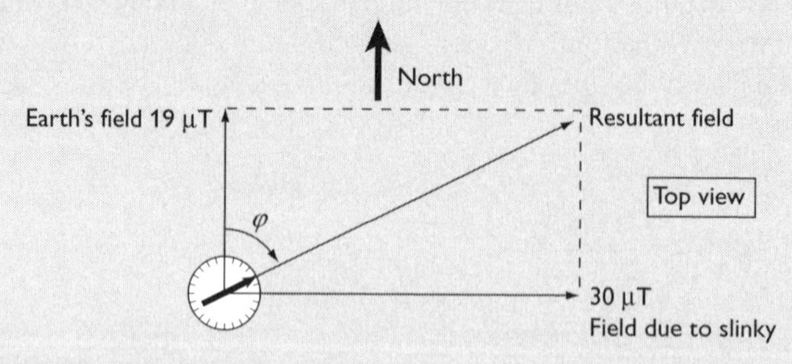

The compass will initially point north and then, when the current is switched on, it will point in the direction of the resultant field.

From the diagram we have $\tan \varphi = \dfrac{30\ \mu\text{T}}{19\ \mu\text{T}}$ so $\varphi = 58°$

You should know how the field in a solenoid and near a straight wire can be investigated experimentally using a Hall probe. You can assume that the probe is pre-calibrated so that it gives the value of the magnetic field directly.

The key point is the orientation of the probe. This must be such that the flat face of the probe is *normal* to the field at the point where the field is being measured. This is shown in the diagrams below.

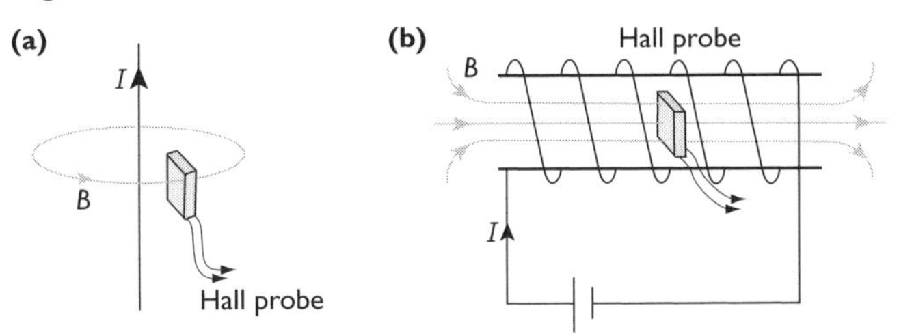

Once the correct positioning of the probe has been established, it is just a question of reading off the value of the field and then measuring the field at different positions, e.g. at varying distances r away from a long wire.

Electromagnetic induction

Magnetic flux

The strength of a magnetic field, its **magnetic flux density**, can be represented by the closeness or 'density' of the lines of force. **Magnetic flux** can be thought of as the *total* number of lines of force in any particular area. We give magnetic flux the symbol Φ and define it as:

$$\Phi = BA$$

where A is the area *normal* to a field of magnetic flux density B. The unit of Φ is T m^2, which is called a weber (Wb).

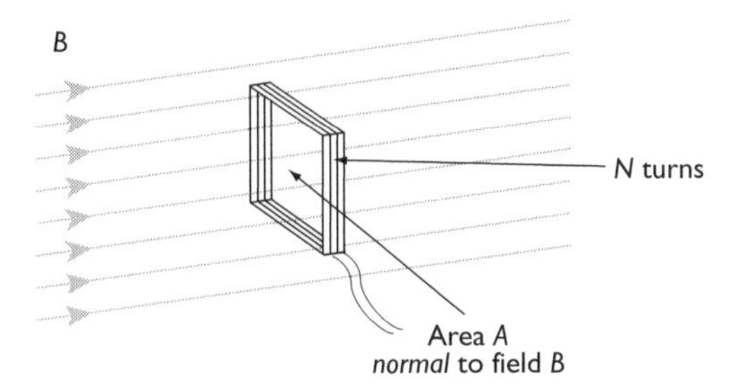

If a magnetic flux Φ passes through a *coil* having N turns, then the turns of the coil are said to be *linked* to the flux and we define the **magnetic flux linkage** as $N\Phi = NBA$. The units are simply weber-turns (Wb-t).

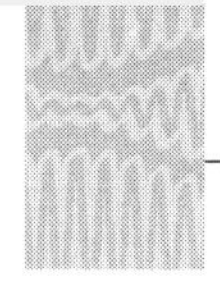

The laws of electromagnetic induction

Electromagnetic induction can be demonstrated experimentally using the simple arrangements shown below.

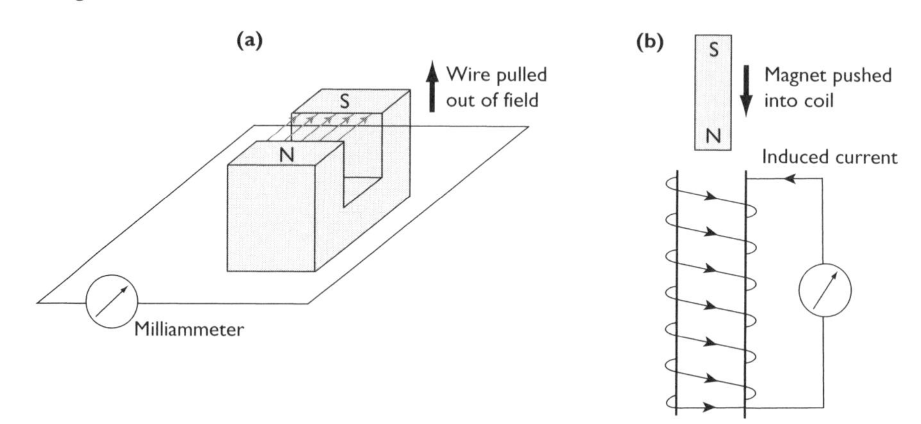

If the wire is moved quickly out of the magnetic field between the poles of the magnet, the meter is seen to flick momentarily and then return to zero. If the wire is pushed back into the field, the meter again flicks momentarily, but in the opposite direction.

Similarly, when the bar magnet is pushed into the coil, the meter briefly registers a current, and when the magnet is removed, a current is observed in the opposite direction. When the magnet is stationary inside the coil, *no* current is recorded. If the magnet is moved in or out more rapidly, the '*induced*' current is seen to be larger.

From the ammeter we can work out the *direction* of the induced current, which always *opposes* the movement of the magnet. If the north pole of the magnet is pushed into the coil, the current induced in the coil is in an anticlockwise direction, creating a north pole at the end of the coil to oppose the magnet. Conversely, when the north pole is pulled out, the current induced in the coil is clockwise, creating a south pole at the end of the coil, which tries to pull the magnet back in again.

Faraday's law is concerned with the magnitude of the *emf* induced in a circuit when there is a change of magnetic flux and **Lenz's law** relates to its direction. Note that it is an *emf* that is induced — a current will be created only if there is a complete electrical circuit for the charge to flow round.

- Faraday's law states that whenever there is a change in flux linking a conductor, an *emf* is induced which is *proportional to the rate of change of flux linkage*.
- Lenz's law states that the induced emf always acts in a direction such that if there is a current in the circuit, it will produce a magnetic field which *opposes the change in magnetic flux*.

The two laws give rise to the following equation for an induced emf $\mathcal{E}$:

$$\mathcal{E} = -N\frac{\Delta\Phi}{\Delta t}$$

The negative sign indicates Lenz's law, which is also an illustration of the **law of conservation of energy**. Imagine pushing a magnet into a horizontal coil: if the induced field were in a direction such that it *attracted* the magnet, we could let go of the magnet and it would accelerate into the coil with ever-increasing acceleration. This would be getting energy for nothing, which would clearly violate the law of conservation of energy.

Worked example

A bar magnet is suspended above a vertical coil of wire. It is then displaced to one side and released such that it oscillates above the coil, as shown in the diagram. The coil of wire has its ends connected to an oscilloscope.

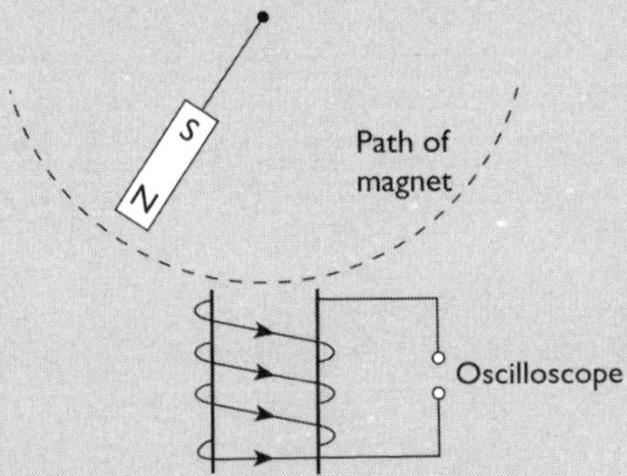

(a) Explain why an emf is induced across the ends of the coil.

(b) By considering Lenz's law, label with an X on the diagram each position of the magnet at which the induced emf changes polarity.

(c) The maximum induced emf is 3.0 mV. Calculate the rate of change of flux needed to induce this emf in a coil of 500 turns.

(d) State three changes that could be made to the apparatus in order to increase the maximum induced emf.

Answer

(a) As the magnet moves across the coil, the magnetic flux cutting the coil will change and so, by Faraday's law, an emf will be induced in the coil. This emf will be proportional to the rate of change of flux.

(b) As the magnet swings down from the position shown, by Lenz's law the induced emf will cause a north pole to be induced in the coil to oppose the approach of the magnet. As the magnet passes over the coil, the emf will change direction so that a south pole is induced in the coil to try to pull the magnet back again. When the magnet gets to the end of its swing and starts to move back towards the coil, the emf will once again change direction so that a north pole is induced to oppose the approaching magnet. The emf thus changes direction at the *centre* and at *each end* of the oscillations.

(c) From $\mathcal{E} = -N\dfrac{\Delta\Phi}{\Delta t}$ we get $= \dfrac{\Delta\Phi}{\Delta t} = \dfrac{\mathcal{E}}{N} = \dfrac{3.0 \times 10^{-3}\ \text{V}}{500} = 6.0\ \mu\text{T m}^2\ \text{s}^{-1}\ (\mu\text{Wb s}^{-1})$

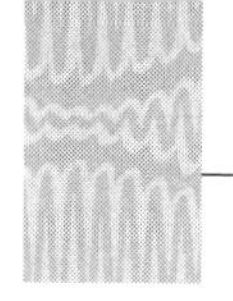

(d) The maximum induced emf could be increased by:
- increasing the amplitude of the oscillations of the magnet
- reducing the length of the suspension
- having a coil with more turns
- putting an iron core inside the coil
- using a stronger magnet

When a *solid* conductor, such as a disc (rather than a coil), moves in a magnetic field, the induced emf causes current loops within the conductor. These are called **eddy currents** (like eddies swirling about in a rock-strewn stream). By Lenz's law, the eddy currents flow in a direction to create magnetic fields that oppose the motion of the disc, producing **eddy current damping**. The ordered kinetic energy of the disc is converted into random kinetic energy of its molecules, i.e. thermal energy. The disc will therefore heat up. This can be a problem in some electrical devices such as the **transformer** (see p. 41).

Worked example

The diagram shows two blades of aluminium. Blade A is complete. Blade B has been cut to form a comb.

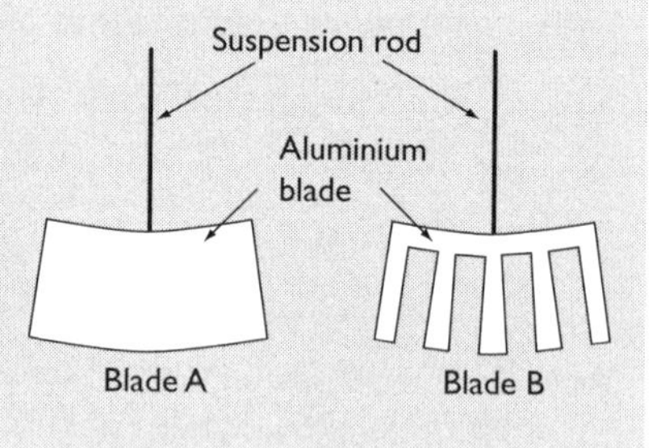

(a) Which electrical property is increased by cutting away the aluminium?

(b) Each blade is suspended in turn between the poles of a strong permanent magnet. Electromagnetic induction produces current loops in blade A as it swings between the poles.

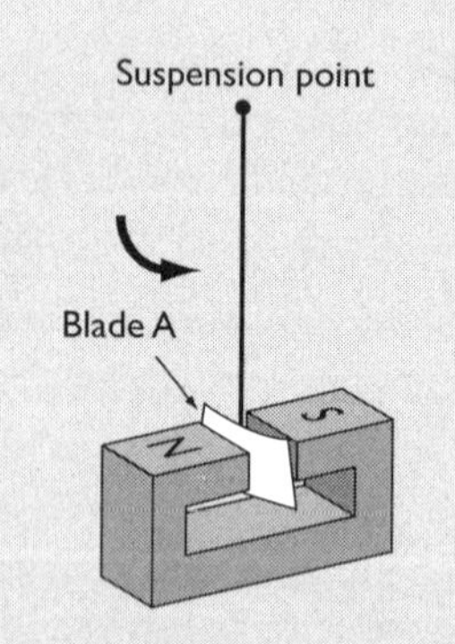

(i) Express Faraday's law of electromagnetic induction in words.

(ii) The oscillations of blade A are rapidly damped. Explain why.

(iii) Suggest why the oscillations of blade B are only lightly damped when it replaces blade A as the blade swinging between the poles.

Answer

(a) Cutting away the aluminium increases the electrical *resistance* of blade B.

(b) (i) Faraday's law states that the induced emf in a conductor is proportional to the rate of change of flux linkage.

(ii) By Lenz's law, the current loops induced in blade A set up a magnetic field that opposes the movement of the blade. As the solid aluminium has a small resistance, these currents are large and so the oscillations are rapidly damped.

(iii) Cutting out pieces of blade B increases its resistance. From $I = V/R$, the induced *currents* are smaller (even though the induced *emf* is the same). The damping is therefore less.

The transformer

One of the main practical applications of electromagnetic induction is the transformer, which is the main component of a battery recharger, such as a mobile-phone recharging unit.

The transformer consists of two coils, the primary and secondary, which in practical transformers are wound on top of each other. The transformer has a laminated, soft iron core to provide a better flux linkage between the two coils. The principle of the transformer is shown below; the coils have been separated for clarity.

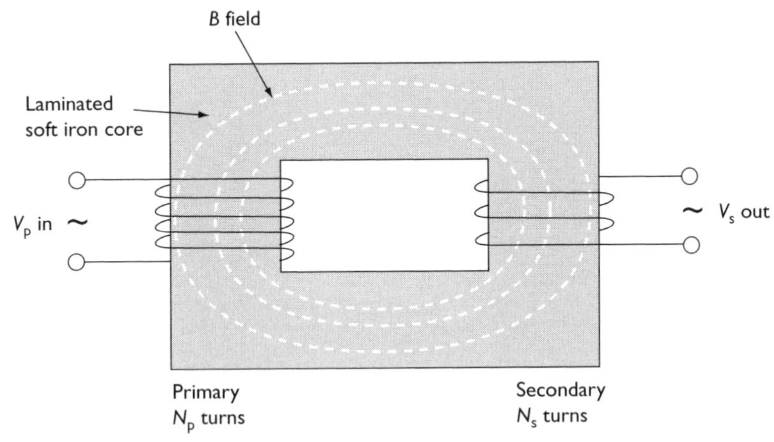

The primary coil, having N_p turns, is connected to a source of *alternating current* of voltage V_p. This produces a *varying* magnetic field in the core, which links with the secondary coil and by electromagnetic induction creates an emf in each of the N_s turns of the secondary coil. For an *ideal transformer*, in which:

- the primary coil has no resistance
- there is no flux leakage, so that there is the same flux through each turn of both the primary and the secondary coils, and
- no current is being taken from the secondary coil

it can be shown that the secondary voltage V_s is given by:

$$\frac{V_p}{V_s} = \frac{N_p}{N_s}$$

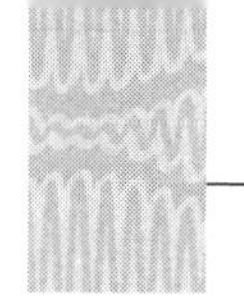

In practice, this relationship is only approximately true for an operational transformer.

- A *step-up* transformer is used to *increase* the voltage, so $N_s > N_p$, which makes $V_s > V_p$.
- For a *step-down* transformer, such as that used in a recharging unit, $N_s < N_p$ so that $V_s < V_p$.

You will have noticed that transformers get warm during use. This is because the changing magnetic field creates eddy currents in the soft iron core. This effect is minimised by making the core from thin iron strips, called laminations, insulated from each other. This increases the electrical resistance of the core and reduces the eddy current heating. Very large transformers, such as those used in the grid system for transmitting electricity across the country, have to be oil-cooled.

Questions & Answers

The following two tests have been compiled using questions from recent past papers. The questions have been chosen to give two balanced papers, with the widest possible coverage of the various topics in the unit. Each paper is worth a total of 40 marks, as in the Unit Test, and so should take you exactly 1 hour. You might like to try to work through a complete paper in the allocated time and then check your answers.

The answers should not be treated as model answers as they represent the bare minimum necessary to get the marks available. The most important reason for studying A-level physics is that you *understand* the subject, not merely learn a set of responses to be repeated parrot-fashion without any thought. In some instances, the difference between an A-grade response and a C-grade response has been suggested. This is not always possible, since many of the questions are short and do not require extended writing.

Examiner's comments

Where appropriate, the answers are followed by examiner's comments, denoted by the icon *e*. They are interspersed in the answers and indicate where credit is due and where lower-grade candidates might typically make common errors. They may also provide useful tips.

Test Paper 1

Question 1

The Earth's gravitational field is radial.

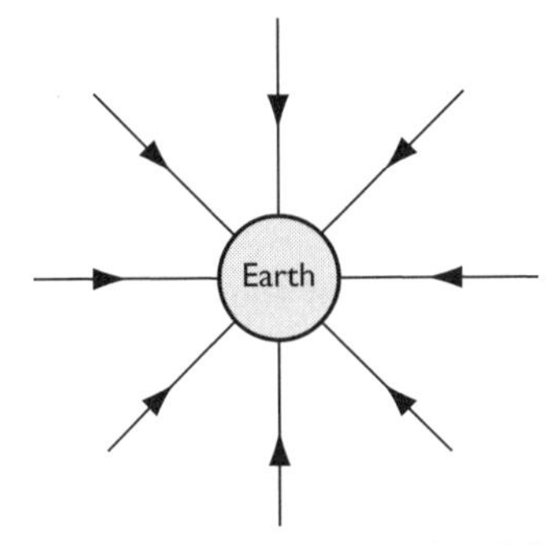

(a) Explain how the diagram indicates that gravitational field strength is a vector quantity. (1 mark)

(b) Add to the diagram three successive equipotential lines. (3 marks)

(c) When a satellite is placed in a circular orbit around the Earth its change in gravitational potential is 2.2×10^7 J kg^{-1}. If the satellite has a mass of 5500 kg, calculate the work done in placing the satellite in this orbit. (2 marks)

(d) Explain why the gravitational potential energy of a satellite does not change as it orbits the Earth. (1 mark)

Total: 7 marks

(January 2003, Question 5)

Answer to Question 1

(a) The field lines have *arrows* on them, which indicate that the gravitational field strength has a *direction* ✓ associated with it, making it a *vector*.

(b)

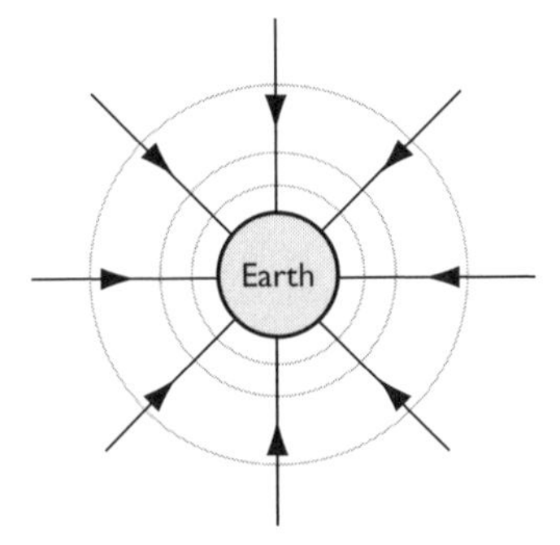

A C-grade candidate might draw *three* ✓ equally spaced *circular* ✓ equipotentials instead of drawing them getting *further apart* ✓.

(c) Work done W = *mass × change in gravitational potential* ✓
$= 5500 \text{ kg} \times 2.2 \times 10^{7} \text{ J kg}^{-1}$
$= 1.2 \times 10^{11} \text{ J}$ ✓

(d) As the satellite orbits the Earth, its distance from the Earth will remain constant. It will therefore move along an equipotential ✓ and so there will be no change in its potential energy.

Question 2

(a) An acetate rod is rubbed with a duster. The rod becomes positively charged. Describe what happens during this process. (2 marks)

(b) The rod is then lowered, at constant speed, towards another positively charged rod that rests on an electronic balance.

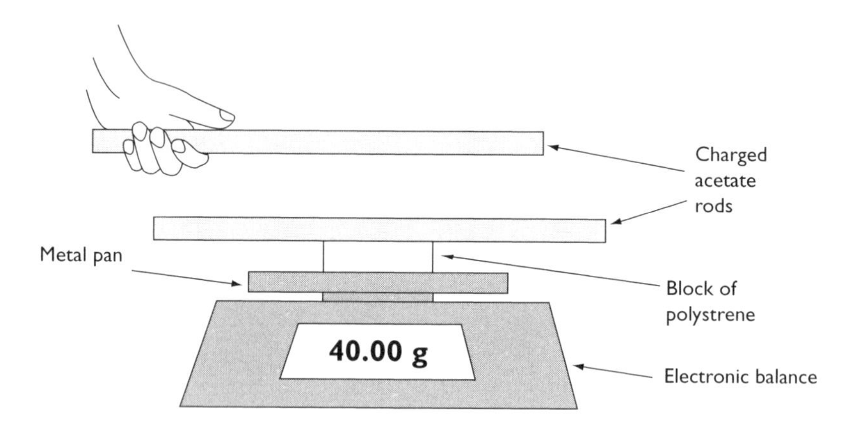

Explain why it is necessary to have the block of polystyrene beneath the bottom rod. (2 marks)

(c) Describe and explain what would happen to the reading on the balance as the top rod slowly approaches, and comes *very close* to, the bottom rod. (4 marks)

Total: 8 marks

(June 2004, Question 1)

Answer to Question 2

(a) When the rod is rubbed, the friction *transfers electrons* ✓ from the rod to the duster. An *equal amount* ✓ of positive charge remains on the rod.

A C-grade candidate may not indicate that there is the *same amount* of charge on both the rod and the duster.

(b) Polystyrene is an *insulator* ✓ and so stops the rod discharging ✓ by preventing the flow of charge through the metal pan.

(c) The balance reading will *increase* ✓ because there is a *force of repulsion* between the *like charges* on the rods ✓. From Coulomb's *inverse square* law ✓ ($F \propto 1/r^2$), the force will increase as the rods get closer and r gets smaller. The reading will increase more rapidly ✓ as the rods become very close and r becomes very small.

Note that the question asks you to describe *and explain*. Just describing what happens would get only 2 marks.

■ ■ ■

Question 3

The diagram shows the path of an electron in a uniform electric field between two parallel conducting plates AB and CD. The electron enters the field at a point midway between A and D. It leaves the field at B.

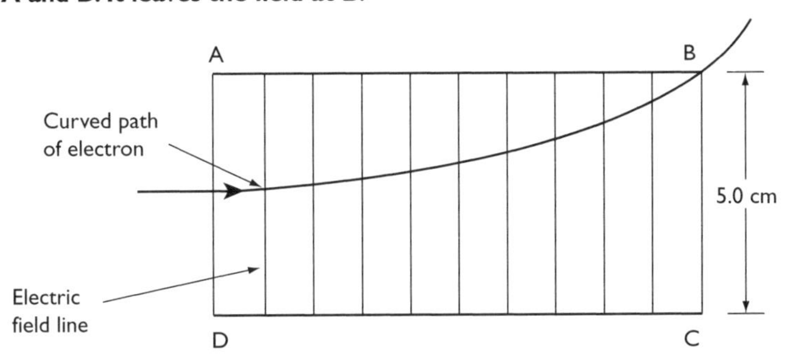

(a) Mark on the diagram the direction of the electric field lines. (1 mark)

(b) (i) The conducting plates are 5.0 cm apart and have a potential difference of 250 V between them. Calculate the force on the electron due to the electric field. (3 marks)

(ii) State the direction of this force on the electron and explain why it does not affect the horizontal velocity of the electron. (2 marks)

(c) To leave the electric field at B the electron must enter the field with a speed of 1.30×10^7 m s^{-1}. Calculate the potential difference required to accelerate an electron from rest to this speed. (3 marks)

(d) A very thin beam of electrons enters a uniform electric field at right angles to the field. The electrons have a range of speeds.

(i) Draw a diagram to show the shape of the beam as it moves through the field.

(ii) On your diagram label which electrons have the fastest speed. (2 marks)

Total: 11 marks

(June 2005, Question 3)

Answer to Question 3

(a) The direction of the field lines is *downwards* ✓.

(b) (i) Electric field $E = \frac{V}{d} = \frac{250\ \text{V}}{5.0 \times 10^{-2}\ \text{m}}$ ✓ $= 5000\ \text{V m}^{-1}$

Force on electron $F = EQ = 5000\ \text{V m}^{-1} \times 1.6 \times 10^{-19}\ \text{C}$ ✓ $= 8.0 \times 10^{-16}\ \text{N}$ ✓

(ii) The force on the electron acts vertically *upwards* ✓. As there is *no component* of this force in the *horizontal* direction ✓, the horizontal velocity is not affected.

(c) Work done on electron = kinetic energy gained by electron

$QV = \frac{1}{2}mv^2$ so $V = \frac{mv^2}{2Q} = \frac{9.11 \times 10^{-31}\ \text{kg} \times (1.3 \times 10^7)^2\ \text{m}^2}{2 \times 1.6 \times 10^{-19}\ \text{C}}$ ✓ ✓

$V = 480\ \text{V}$ ✓

Note that it is often easier to rearrange the equation, as above, before putting in the data. In the examination, data such as the mass of an electron will be given at the end of the question paper. You need to remember to look for it.

(d)

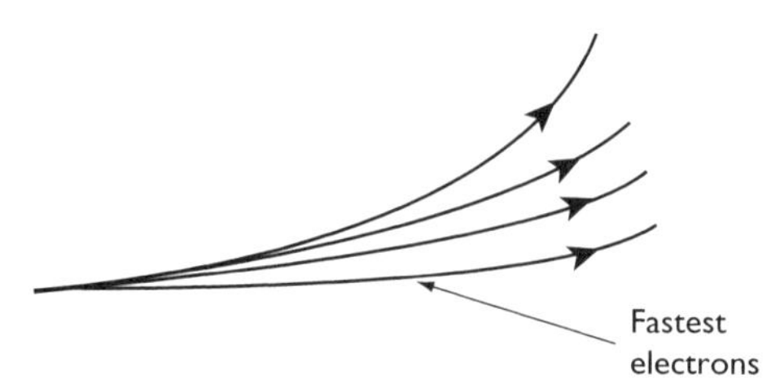

Note the lines should spread out from *one point* ✓, with the fastest electrons deflected *least* ✓.

■ ■ ■

Question 4

(a) The circuit diagram shows a long straight wire LM carrying a current *I* in the direction shown.

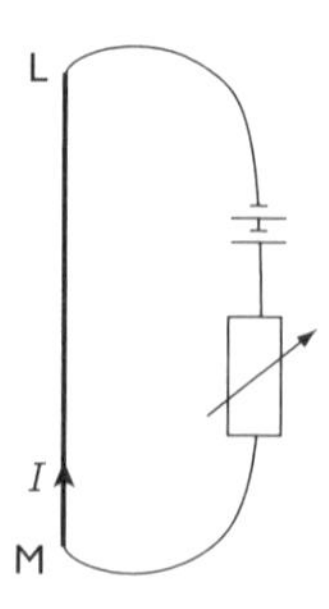

Describe how you would investigate the variation of magnetic flux density *B* with perpendicular distance *r* from LM in a region round the centre of the wire. You may add to the above diagram if you wish. (4 marks)

(b) A typical graph of B against $\frac{1}{r}$ for a wire carrying a current I is shown below.

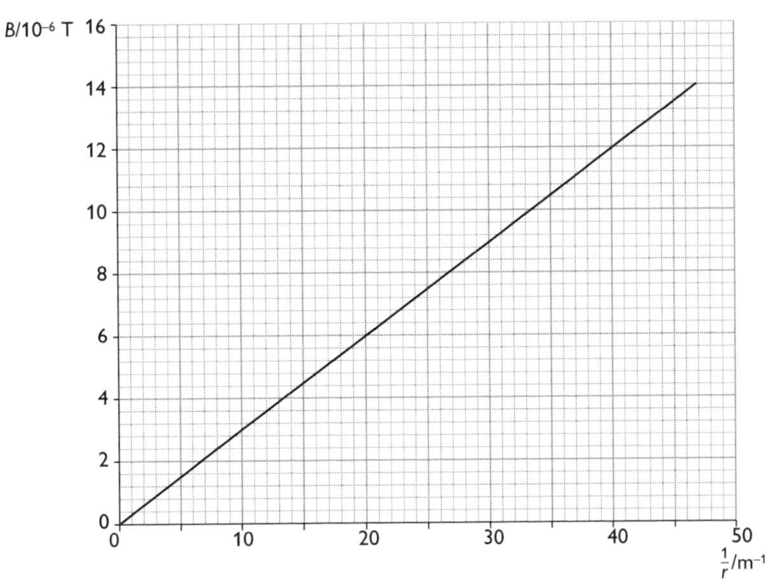

(i) Describe the relationship between B and r shown by the graph. (1 mark)

(ii) Use the graph to determine the value of the current I in the wire. (3 marks)

Total: 8 marks

(June 2005, Question 4)

Answer to Question 4

(a)

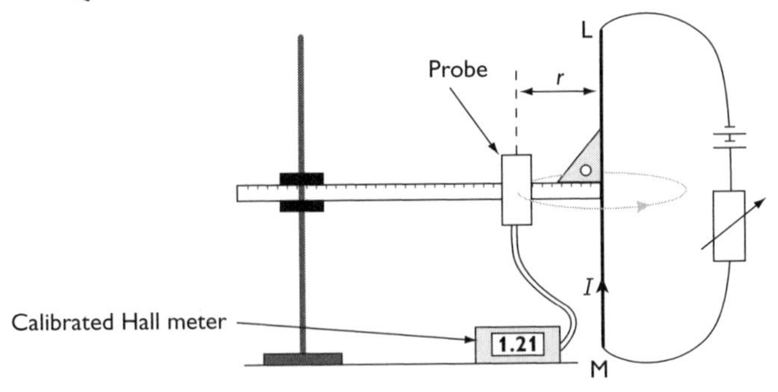

Note that the diagram should show a Hall probe connected to a pre-calibrated meter ✓ and the slice of the probe should be *perpendicular* to the field ✓.

The distance r, *perpendicular* to the wire, can be measured using a rule clamped horizontally and a set square, as shown ✓. The field B is then measured for different values of r ✓.

Care must be taken to draw a good, labelled diagram to show these points, particularly to indicate clearly the correct orientation of the Hall slice.

(b) (i) The graph shows that B is *inversely proportional* to r [or $B \propto 1/r$] ✓.

(ii) From $B = \dfrac{\mu_0 I}{2\pi r}$ we get $I = \dfrac{2\pi Br}{\mu_0}$ where Br = gradient of graph

$$\text{Gradient} = \frac{12 \times 10^{-6}\ \text{T}}{40\ \text{m}^{-1}} = 3.0 \times 10^{-7}\ \text{T m} ✓$$

$$I = \frac{2\pi \times \text{gradient}}{\mu_0} ✓ = \frac{2\pi \times 3.0 \times 10^{-7}\ \text{T m}}{4\pi \times 10^{-7}\ \text{N A}^{-2}} = 1.50\ \text{A} ✓$$

■ ■ ■

Question 5

(a) State Lenz's law of electromagnetic induction. (2 marks)

(b) A bar magnet is dropped from rest through the centre of a coil which is connected to a resistor and datalogger.

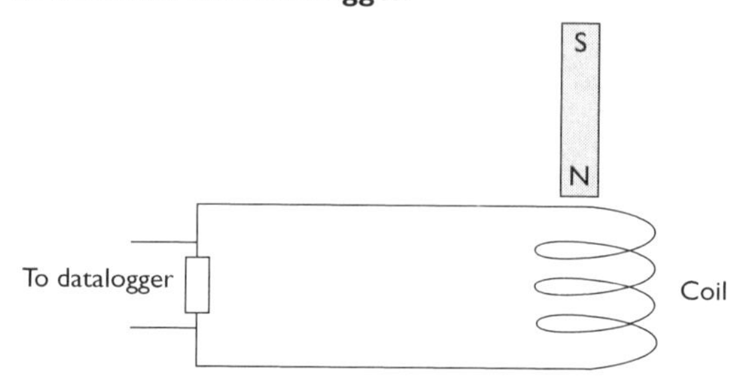

(i) State the induced polarity on the top side of the coil as the magnet falls towards it.

(ii) Add an arrow to the wire to show the direction of the induced current as the magnet falls towards the coil. (2 marks)

(c) The graph shows the variation of induced current in the resistor with time as the magnet falls.

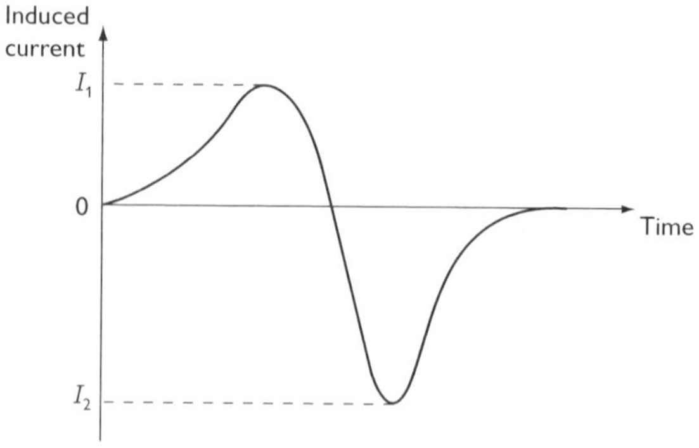

Explain why the magnitude of I_2 is greater than I_1. (2 marks)

Total: 6 marks

(June 2004, Question 4)

Answer to Question 5

(a) Lenz's law states that the *induced emf* always acts in a *direction* ✓ such that if there is a current in the circuit, it will produce a magnetic field which *opposes the change in magnetic flux* ✓.

(b) (i) The top of the coil will become a *north* ✓ pole to oppose the north pole of the falling magnet.

(ii) The current in the circuit is *anticlockwise* ✓.

(c) The magnet is accelerating ✓ under gravity. *The rate of change of flux* is therefore *greater* when the magnet leaves the coil than when it enters ✓. The *induced emf*, and so the induced current, is therefore greater ✓.

e Remember, it is an *emf* that is induced, which will create a current only if there is a complete circuit (as there is in this case).

Test Paper 2

Question 1

(a) State Newton's law for the gravitational force between point masses. (2 marks)

(b) Use this law to show that the gravitational field strength g at a distance r from the centre of the Earth, where r is greater than or equal to the radius R of the Earth, is given by:

$$g = \frac{GM}{r^2}$$

where M is the mass of the Earth. (1 mark)

(c) Use the axes below to plot a graph to show how g varies as the distance r increases from its minimum value of R to a value of $4R$.

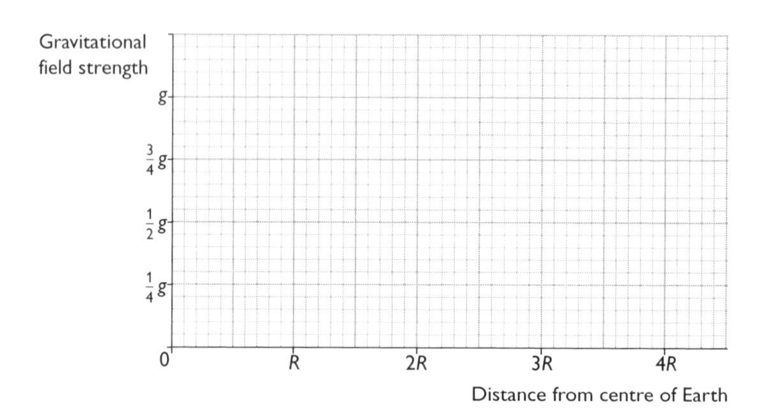

(3 marks)

(June 2005, Question 2) **Total: 6 marks**

Answer to Question 1

(a) Newton's law states that the force between two point masses is proportional to the *product* of the two masses ✓ and is *inversely proportional* to the *square* of the *distance* between the masses ✓.

A C-grade candidate might miss out one of the important *italicised* terms and so lose a mark.

(b) From Newton's law $F = mg = \frac{GMm}{r^2}$ so $g = \frac{GM}{r^2}$ ✓

(c)

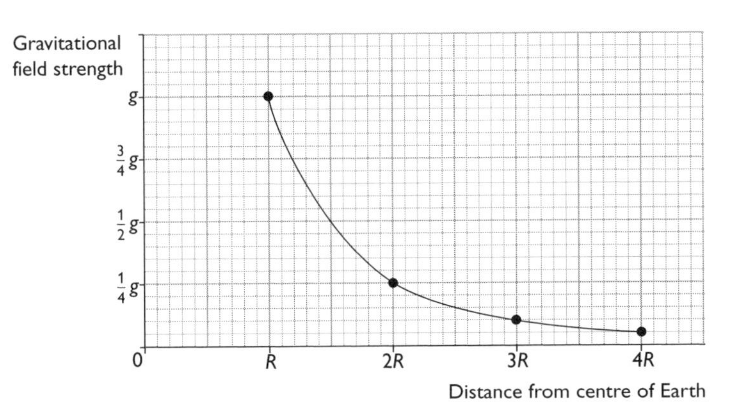

As g is inversely proportional to the *square* of r, the values of g become $\frac{1}{4}g$ at $2R$, $\frac{1}{9}g$ at $3R$ etc.

A C-grade candidate will probably get a downward curve ✓ starting at g when the distance is R ✓, but may not put in at least two more values ✓.

Question 2

The diagram shows a circuit that is to be used to charge a 500 μF capacitor using a battery of negligible internal resistance. The ammeter measures the current during the charging process from the instant that the switch S is closed.

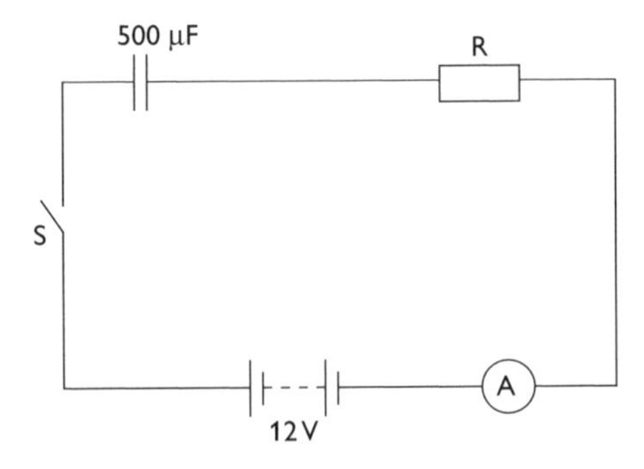

The graph on p. 54 shows how the current *I* in the ammeter varies with time.

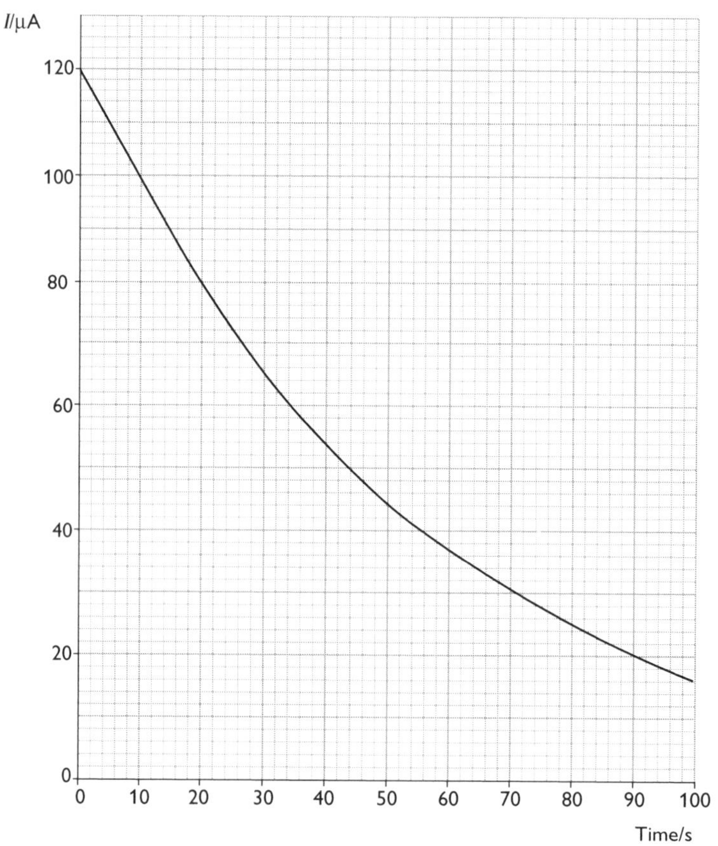

(a) Use the graph to estimate the charge stored during the first 20 s. (3 marks)

(b) The charge stored after 40 s is 3300 μC. Use this value to calculate the potential difference across the capacitor at 40 s. (2 marks)

(c) Determine the potential difference across the resistor R at 40 s. (1 mark)

(d) Hence, or otherwise, calculate the resistance of R. (3 marks)

(e) On the graph, sketch the curve which would be obtained if the 500 μF capacitor were replaced by one of smaller capacitance. (1 mark)

Total: 10 marks

(January 2005, Question 3)

Answer to Question 2

(a) From $Q = It$, the charge is the area under the graph. To a very good approximation, the area up to 20 s is a trapezium ✓, so

$$\text{charge } Q \approx \tfrac{1}{2}(120 + 80)\ \mu\text{A} \times 20\ \text{s}\ ✓ \approx 2000\ \mu\text{C}\ ✓$$

(Alternatively, there are about 20 large squares, each ≡ 100 μC so $Q \approx 2000$ μC)

Don't spend a long time counting small squares — look for quick and easy approximations.

(b) PD across capacitor $V = \frac{Q}{C} = \frac{3300\ \mu C}{500\ \mu F}$ ✓ = 6.6 V ✓

(c) PD across resistance = 12 V (battery) – 6.6 V (across capacitor) = 5.4 V ✓

(d) At 40 s, from graph, current I = 54 μA ✓

$R = \frac{V}{I} = \frac{5.4\ V}{54\ \mu A}$ ✓ = 100 000 Ω ✓ = 100 kΩ

[Alternatively, at the point when the switch is closed (t = 0), I = 120 μA ✓ and the full 12 V of the battery will be across the resistor, giving

$R = \frac{V}{I} = \frac{12\ V}{120\ \mu A}$ ✓ = 100 000 Ω ✓ = 100 kΩ]

(e) The curve should start at 120 μA and be drawn *below* the given curve ✓.

Question 3

The diagram shows a rigid wire clamped at L and M so that it cannot move. The clamps are not shown. A U-shaped magnet provides a uniform, horizontal magnetic field between its pole pieces. The horizontal rigid wire is perpendicular to this field. The magnet rests on the pan of the balance.

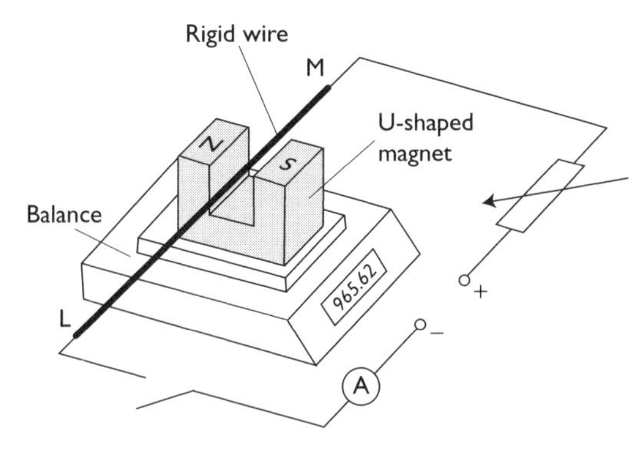

With the switch open the balance reads 965.62 g. The switch is closed and the reading changes to 966.07 g. The current in the wire is 1.5 A.

(a) (i) Explain why the balance reading increases. You may be awarded a mark for the clarity of your answer. (4 marks)

(ii) The length of the U-shaped magnet is 6.0 cm. Assuming the magnetic field is zero outside the pole pieces, calculate the flux density of the uniform, horizontal magnetic field between them. (3 marks)

(b) A student proposes to use this apparatus to show the relationship between the electromagnetic force produced and the current in the rigid wire.

(i) Describe the procedure he should follow. (2 marks)

(ii) Sketch a graph of the expected results using the axes below. Label the axes.

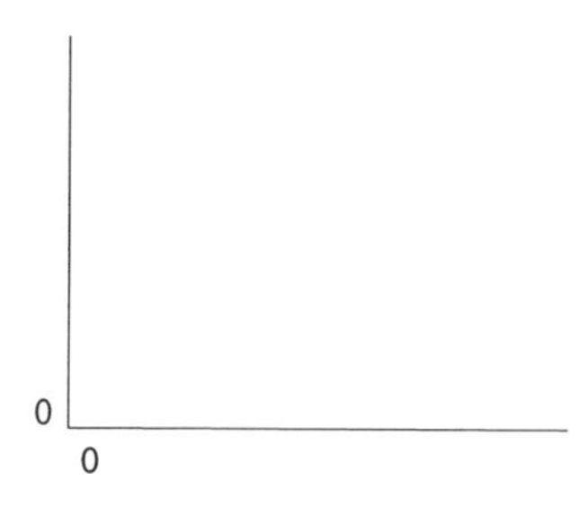

(1 mark)

(January 2006, Question 4) **Total: 10 marks**

Answer to Question 3

(a) (i) The current in the rigid wire experiences a *force* due to the magnetic field ✓. By Fleming's left-hand rule this force is *upwards* ✓. By Newton's third law the wire exerts an equal and opposite *downward* force on the magnet ✓.

To get the 'clarity' mark ✓, each step of your argument should be clearly explained, using the correct technical language of physics.

(ii) $F = \Delta mg = (966.07 - 965.62) \times 10^{-3}\text{ kg} \times 9.8\text{ N kg}^{-1}$ ✓

$F = BIl$ so $B = \dfrac{F}{Il} = \dfrac{0.45 \times 10^{-3}\text{ kg} \times 9.81\text{ N kg}^{-1}}{1.5\text{ A} \times 6.0 \times 10^{-2}\text{ m}}$ ✓ $= 4.9 \times 10^{-2}\text{ T}$ ✓

You need to be careful with units here, e.g. converting Δm to kg and 6.0 cm to m.

(b) (i) Close the switch and note the ammeter and balance readings ✓. Adjust the variable resistor to vary the current and record a range of values of the current I and balance readings ✓. Calculate the corresponding forces from $F = \Delta mg$.

A C-grade candidate may not describe *how* the current would be altered.

(ii) Plot a graph of F against I:

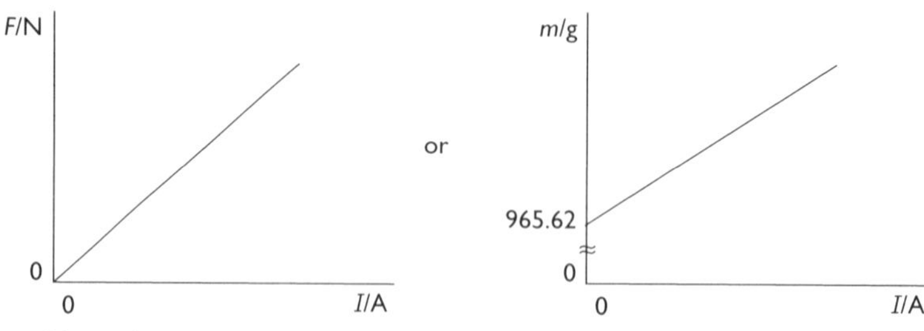

You could simply plot a graph of the balance reading m against I, in which case the line would not go through the origin, but would cut the m-axis at 965.62 g.

Question 4

(a) A metal 'slinky' spring can be used as a long solenoid. It is stretched out so that each turn is 1.6 cm from its nearest neighbour. A current of 0.50 A is passed through the coils of the solenoid.

(i) Calculate the number of turns per metre in the solenoid. (1 mark)

(ii) Show that the magnetic field strength in the middle of the solenoid is about 4×10^{-5} T. (2 marks)

(b) A longitudinal compression pulse is sent along the solenoid. The resulting changes in the magnetic field strength are detected by a Hall probe fixed inside the solenoid. As the compression pulse passes the Hall probe, the reading changes as follows:

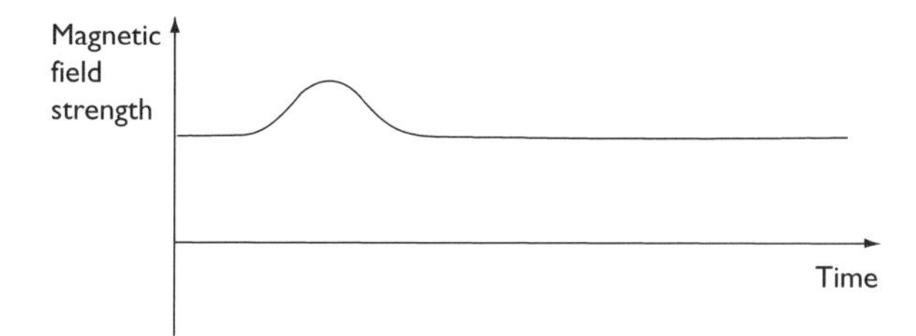

Explain the shape of the graph. (2 marks)

(c) The solenoid is replaced by a heavier one that has an identical number of turns per unit length. A rarefaction pulse is then sent along this heavier solenoid. This pulse travels at a slower speed than the pulse in the first solenoid.
Show on the axes above how the reading of the Hall probe will now vary. (2 marks)

Total: 7 marks

(January 2003, Question 4)

Answer to Question 4

(a) (i) $n = \dfrac{1}{1.6 \times 10^{-2}\ \text{m}} = 62.5\ \text{m}^{-1}$ ✓

(ii) $B = \mu_0 nI = 4\pi \times 10^{-7}\ \text{N A}^{-2} \times 62.5\ \text{m}^{-1} \times 0.50\ \text{A}$ ✓

$= 3.9 \times 10^{-5}$ T ✓ $\approx 4 \times 10^{-5}$ T

(b) When the compression reaches the Hall probe, the magnetic field strength increases because the turns in the compression are closer together: *n* increases ✓. The field reaches a maximum when the centre of the compression is in line with the Hall probe and then starts to decrease again as the compression moves away from the probe ✓.

(c)

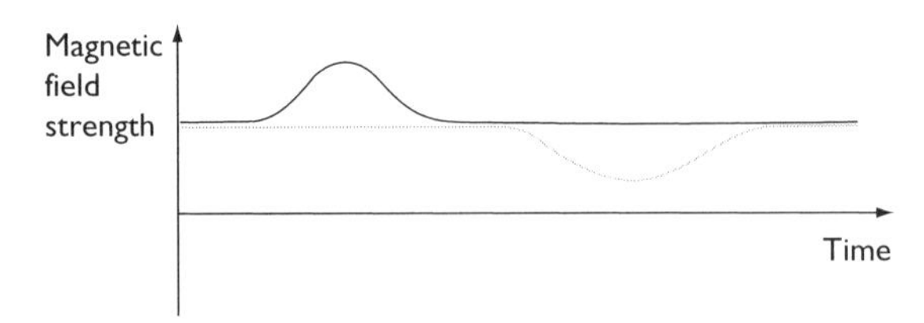

C-grade candidates may not realise that as this is a *rarefaction* the coils will be *further apart*, thus *reducing n* and making the field *smaller* ✓. The pulse will also take longer to reach the Hall probe and be more spread out as it is travelling more slowly ✓.

Question 5

(a) Explain the action of the transformer shown below.

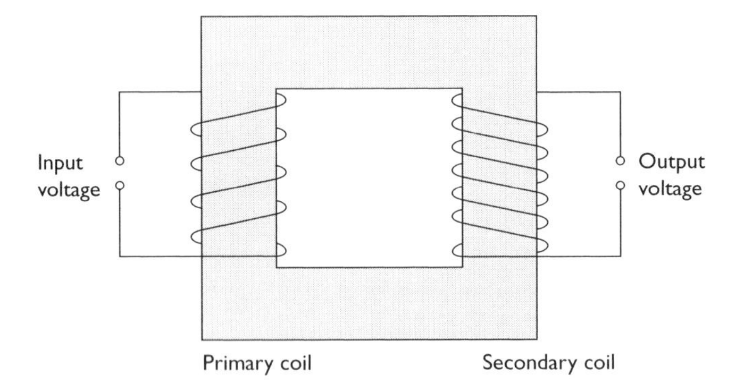

(5 marks)

(b) The following diagram shows an ideal step-down transformer.

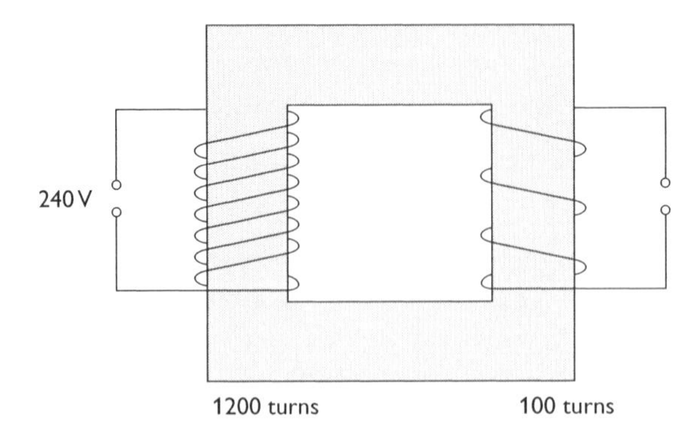

The input voltage is 240 V. Calculate the output voltage of the transformer. (2 marks)

Total: 7 marks

(January 2003, Question 3)

Answer to Question 5

(a) If a source of *alternating* voltage ✓ is connected to the input, the alternating current in the primary coil will create an alternating magnetic *flux* in the soft iron core ✓. This *changing* flux will link with the turns in the secondary coil ✓ and *induce* ✓ an alternating emf in the secondary coil. As there are more turns in the secondary coil than in the primary coil, the output voltage will be *greater* than the input voltage ✓.

The important points are *italicised*. Note that you must state that the input is *alternating*. As the question refers to 'the transformer shown', you should explain that it is a 'step-up' transformer. These points might well be missed by C-grade candidates.

(b) From $\frac{V_P}{V_s} = \frac{N_P}{N_s}$ we get $\frac{240\ \text{V}}{V_s} = \frac{1200}{100}$ ✓ $= 12$

$$V_s = \frac{240\ \text{V}}{12} = 20\ V \checkmark$$